Collins

INTERNATIONAL PRIMARY MATHS

Workbook 5

William Collins' dream of knowledge for all began with the publication of his first book in 1819. A self-educated mill worker, he not only enriched millions of lives, but also founded a flourishing publishing house. Today, staying true to this spirit, Collins books are packed with inspiration, innovation and practical expertise. They place you at the centre of a world of possibility and give you exactly what you need to explore it.

Collins. Freedom to teach.

Published by Collins
An imprint of HarperCollinsPublishers
The News Building
1 London Bridge Street
London
SE1 9GF

HarperCollinsPublishers
Macken House,
39/40 Mayor Street Upper,
Dublin 1
D01 C9W8
Ireland

Browse the complete Collins catalogue at
www.collins.co.uk

© HarperCollinsPublishers Limited 2021

10 9 8 7 6 5 4

ISBN 978-0-00-836949-1

British Library Cataloguing-in-Publication Data
A catalogue record for this publication is available from the British Library.

Author: Paul Hodge
Series editor: Peter Clarke
Publisher: Elaine Higgleton
Product developer: Holly Woolnough
Project manager: Mike Harman (Life Lines Editorial Services)
Development editor: Joan Miller
Copyeditor: Catherine Dakin
Proofreader: Tanya Solomons
Cover designer: Gordon MacGilp
Cover illustrator: Ann Paganuzzi
Typesetter: Ken Vail Graphic Design
Illustrators: Ann Paganuzzi and Ken Vail Graphic Design
Production controller: Lyndsey Rogers
Printed and bound by Grafica Veneta S. P. A.

With thanks to the following teachers and schools for reviewing materials in development: Antara Banerjee, Calcutta International School; Hawar International School; Melissa Brobst, International School of Budapest; Rafaella Alexandrou, Pascal Primary Lefkosia; Maria Biglikoudi, Georgia Keravnou, Sotiria Leonidou and Niki Tzorzis, Pascal Primary School Lemessos; Taman Rama Intercultural School, Bali.

MIX
Paper from
responsible sources
FSC www.fsc.org **FSC™ C007454**

This book is produced from independently certified FSC™ paper to ensure responsible forest management.

For more information visit: **www.harpercollins.co.uk/green**

The publishers gratefully acknowledge the permission granted to reproduce the copyright material in this book. Every effort has been made to trace copyright holders and to obtain their permission for the use of copyright material. The publishers will gladly receive any information enabling them to rectify any error or omission at the first opportunity.

Contents

Number

1 Counting and sequences   6
2 Addition of whole numbers   14
3 Subtraction of whole numbers   22
4 Multiples, factors, divisibility, primes and squares   30
5 Whole number calculations   38
6 Multiplication of whole numbers (A)   46
7 Multiplication of whole numbers (B)   54
8 Division of whole numbers (A)   62
9 Division of whole numbers (B)   70
10 Place value and ordering decimals   78
11 Place value, ordering and rounding decimals   86
12 Fractions (A)   94
13 Fractions (B)   102
14 Percentages   110
15 Addition and subtraction of decimals   118
16 Multiplication of decimals   126
17 Fractions, decimals and percentages   134
18 Proportion and ratio   142

Geometry and Measure

19 Time   150
20 2D shapes and symmetry   158
21 3D shapes   166
22 Angles   174
23 Perimeter and area   182
24 Coordinates   190
25 Translation and reflection   198

Statistics and Probability

26 Statistics   206
27 Statistics and probability   214

How to use this book

This book is used during the middle part of a lesson when it is time for you to practise the mathematical ideas you have just been taught.

- An **objective** explains what you should know, or be able to do, by the end of the lesson.

You will need
- Lists the resources you need to use to answer some of the questions.

There are two pages of practice questions for each lesson, with three different types of question.

1 Some question numbers are written on a **circle**. These questions may be **easier**. They may also practise mathematical ideas you have learned before. These questions will help you answer the rest of the questions on the two pages.

3 Some question numbers are written on a **triangle**. These questions provide **practice** on mathematical ideas you have just been taught. They help you to understand the ideas better.

5 Some question numbers are written on a **square**. These questions are slightly more **challenging**. They make you think more deeply about the mathematical ideas.

You won't always have to answer all the questions on the two pages. Your teacher will tell you which questions to answer.

HINT

Draw a ring around the question numbers your teacher tells you to answer.

 Questions with a star beside them require you to Think and Work Mathematically (TWM). You might want to use the TWM star at the back of the Student's Book to help you.

Date: _____

At the bottom of the second page there is room to write the date you completed the work on these pages. If it took you longer than one day, write all of the dates you worked on these pages.

Self-assessment

Once you have answered the questions on the pages, think carefully about how easy or hard you find the ideas. Circle the face that describes you best.

 I can do this.

 I'm getting there.

 I need some help.

Number

Lesson 1: **Counting on and back (1)**

• Count on and count back in steps of 7, 8 or 9

1 Count on or back in the steps given.

a Count on in steps of 7.

14, ☐, ☐, ☐, 42, ☐, ☐, 63, ☐, ☐

b Count back in steps of 9.

99, ☐, ☐, ☐, 63, ☐, ☐, 36, ☐, ☐

c Count on in steps of 8.

32, ☐, ☐, ☐, 64, ☐, ☐, 88, ☐, ☐

d Count back in steps of 7.

84, ☐, ☐, ☐, 56, ☐, ☐, 35, ☐, ☐

2 Count on or back in the steps given.

a Count on in steps of 7.

2, ☐, ☐, ☐, 30, ☐, ☐, 51, ☐, ☐

b Count back in steps of 9.

95, ☐, ☐, ☐, 59, ☐, ☐, 32, ☐, ☐

c Count on in steps of 8.

213, ☐, ☐, ☐, 245, ☐, ☐, 269, ☐, ☐

d Count back in steps of 7.

505, ☐, ☐, ☐, 477, ☐, ☐, 456, ☐, ☐

e Count on in steps of 9.

787, ☐, ☐, ☐, 823, ☐, ☐, 850, ☐, ☐

Number

 3 Draw a ring around the numbers that you would say as part of the counting sequence.

a Count forward in 8s from 30.

38   44   46   54   60   62   70   77   79   84   86   90   92   94

b Count back in 7s from 151.

145   144   137   132   131   123   116   108   101   95   94   90   88   81

c Count forward in 9s from 777.

785   794   796   805   804   812   823   831   832   839   847   858   866   867

4 A sack of potatoes has a mass of 144 kg. Five small bags of potatoes, each with a mass of 8 kg, are emptied into the sack. What is the mass of the sack now? ☐ kg. Draw a number line to show your answer.

5 Write the terms for each sequence.

a Count forward in 9s from 8.

2nd term ☐   3rd term ☐   5th term ☐

b Count back in 7s from 60.

2nd term ☐   4th term ☐   5th term ☐

c Count forward in 8s from 21.

3rd term ☐   4th term ☐   6th term ☐

d Count back in 9s from 80.

3rd term ☐   4th term ☐   6th term ☐

Date: _____

Lesson 2: **Counting on and back (2)**

Number

> • Count on and count back in steps of a constant size from different numbers, including negative numbers

1 Write the numbers in the sequence. Continue the count as far as you can on each number line.

a Count back in 4s.

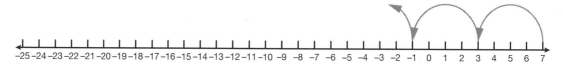

Sequence: 7, 3, −1, _____

b Count forward in 3s.

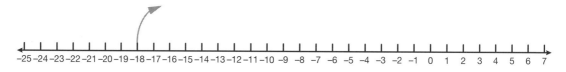

Sequence: −18, _____

c Count back in 5s.

Sequence: 4, _____

2 Write the missing numbers in the temperature table.

Start temp. (°C)	15	−19	20	−14
Temp. rise/fall	**Fall** 4 degrees per hour	**Rise** 5 degrees per hour	**Fall** 7 degrees per hour	**Rise** 8 degrees per hour
Number of hours	6	7	5	4
End temp. (°C)				

3 A diver's work involves diving below the surface of a lake.

Work out the position of the diver at the end of each journey.

Write your answers as metres (m) above or below sea level.

Positive numbers indicate a position above sea level.

Negative numbers indicate a position below sea level.

Starting position (metres above/ below sea level)	Journey A	Journey B	End position (metres above/ below sea level)
15 m	Dive of 7 m a minute for 3 minutes	Climb of 8 m a minute for 2 minutes	
−23 m	Climb of 9 m a minute for 4 minutes	Dive of 8 m a minute for 5 minutes	

4 A number track is marked out in numbers from −60 to 40. Start and end numbers are chosen and, between them, the track is divided into sections of equal length. Write the missing numbers in the table.

Start number	End number	Number of equal sections	Value of each section
−15	25	5	
−6	30	4	
−13	29	7	
−31	33		8
−44	37		9
−55	17		6

Date: _____

Number

Lesson 3: **Number sequences**

- Find missing terms in a sequence

1 Use the rule for each sequence to find the missing terms.

a Rule: add 8

89	97			121		137			161

b Rule: add 9

114	123		141			168	177		

c Rule: add 7

285		299			320		334	341	

d Rule: add 11

483	494			527		549	560		

2 Each of the sequences below is linear. Write in the missing terms.

a | 38 | | | | 58 |

b | 77 | | | | | 97 |

c | 152 | | | | | 192 |

d | 26 | | | | 74 |

e | 225 | | | | | | 279 |

f | 75 | | | | 43 |

g | 101 | | | | 65 |

h | 125 | | | | | 90 |

> **Remember!**
>
> A linear sequence increases or decreases by the same amount each time.

Number

3 Solve the word problems.

a The masses of four crates of apples form a linear sequence. The first crate has a mass of 466 g and the fourth a mass of 502 g. What are the masses of the other two crates?

[ ] g and [ ] g [_____]

b The volumes of orange juice in five jugs form a linear sequence. The first jug holds 227 ml of juice and the fifth holds 255 ml. How much juice do the other jugs hold?

[ ] ml, [ ] ml and [ ] ml

[_____]

c The heights of eight sunflowers planted in a row form a linear sequence. The first sunflower is 673 cm and the eighth is 729 cm. What are the heights of the other six sunflowers?

[ ] cm, [ ] cm, [ ] cm, [ ] cm, [ ] cm and [ ] cm

[_____]

4 Each of the sequences below is linear. Write in the missing terms.

a [ ] [ ] 498 [ ] [ ] [ ] 546 [ ] [ ] [ ]

b [ ] 651 [ ] [ ] [ ] 711 [ ] [ ] [ ] [ ]

c [ ] [ ] [ ] 449 [ ] [ ] [ ] [ ] 574 [ ]

d [ ] [ ] [ ] 407 [ ] [ ] [ ] 479 [ ] [ ]

Date: _____

Number

Lesson 4: **Square and triangular numbers**

- Understand square and triangular numbers and extend sequences of these numbers

You will need

- Resource sheet 2: Triangular dot paper
- ruler
- coloured pencils: red, yellow

1 Draw a ring around the pictures that **do not** represent a square number.

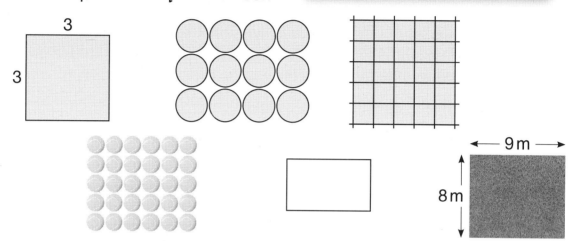

2 Draw the first six triangular numbers. If you run out of room, use Resource sheet 2 to continue.

3 Colour the square numbers red. Draw a ring around the triangular numbers yellow.

Which numbers are both square and triangular?

1	2	3	4	5	6	7	8	9	10
11	12	13	14	15	16	17	18	19	20
21	22	23	24	25	26	27	28	29	30
31	32	33	34	35	36	37	38	39	40
41	42	43	44	45	46	47	48	49	50
51	52	53	54	55	56	57	58	59	60
61	62	63	64	65	66	67	68	69	70
71	72	73	74	75	76	77	78	79	80
81	82	83	84	85	86	87	88	89	90
91	92	93	94	95	96	97	98	99	100

Number

4 Lucía arranges counters to form triangular numbers. She finds she needs 6 counters for the 3rd term, 10 counters for the 4th term and 15 counters for the 5th term.

How many counters will she need for the 9th term?

5 Investigate this statement:

A triangular number can never end in 2, 4, 7 or 9.

Is this statement true or false? Write your working in the box below.

Date: _____

13

Number

Lesson 1: **Adding 2- and 3-digit numbers**

- Estimate and add 2- and 3-digit numbers, choosing the best mental or written method

You will need
- squared paper
- ruler

1 Choose an appropriate strategy to solve each calculation.

a $73 + 60 =$ ☐

b $133 + 80 =$ ☐

c $267 + 60 =$ ☐

d $446 + 120 =$ ☐

e $524 + 170 =$ ☐

f $739 + 250 =$ ☐

g $456 + 750 =$ ☐

h $783 + 440 =$ ☐

 2 Use the most appropriate strategy to find each sum. Estimate the answer first. Use squared paper if you need to.

a $34 + 281 + 66 =$ ☐

Estimate: ☐

b $172 + 89 + 133 =$ ☐

Estimate: ☐

c $352 + 77 + 273 =$ ☐

Estimate: ☐

d $94 + 367 + 325 =$ ☐

Estimate: ☐

e $463 + 146 + 236 =$ ☐

Estimate: ☐

f $384 + 478 + 247 =$ ☐

Estimate: ☐

g $754 + 83 + 758 =$ ☐

Estimate: ☐

h $867 + 759 + 936 =$ ☐

Estimate: ☐

 3 Write the letter of each calculation in the correct part of the Carroll diagram. Don't work out the answers – use estimation.

a $346 + 247 + 406$

b $548 + 234 + 219$

c $388 + 364 + 247$

d $197 + 285 + 519$

e $434 + 435 + 128$

f $555 + 97 + 346$

g 362 + 247 + 276 + 118   **h** 318 + 273 + 198 + 209

i 544 + 87 + 237 + 134   **j** 436 + 373 + 85 + 108

3-digit answers	Not 3-digit answers

4 Li pours three jugs of orange juice with 476 ml, 787 ml and 538 ml into container A. Three jugs of blackcurrant are poured into container B: 647 ml, 743 ml and 968 ml. Calculate the amount of juice in each container. How much more juice does container B hold than container A?

Container A: [ ] ml   Container B: [ ] ml

Difference: [ ] ml

5 The addends for these sums have been mixed up. Write the letter codes for the three addends that make each sum. Write your estimates in pencil first.

878 = [ ] + [ ] + [ ]

680 = [ ] + [ ] + [ ]

706 = [ ] + [ ] + [ ]

A	347
B	88
C	284
D	116
E	476
F	217
G	75
H	514
I	147

Date: _____

Number

Lesson 2: **Adding 4-digit numbers**

Number

- Estimate and add 4-digit numbers using the formal written method

You will need
- squared paper
- ruler

1 Complete each calculation.

a
```
    4 2 3
+   3 4 5
```

b
```
    7 6 2
+   1 8 4
```

c
```
    4 2 7
+   5 3 7
```

d
```
    5 4 5 6
+   2 4 3 2
```

e
```
    6 3 4 4
+   1 2 7 4
```

f
```
    6 2 3 7
+   4 5 4 7
```

2 Use the formal written method to calculate each sum. Estimate the answer first. Use squared paper to show your working.

a   4526 + 2283 =

Estimate:

b   3744 + 4238 =

Estimate:

c   5267 + 1921 =

Estimate:

d   6236 + 2815 =

Estimate:

e   7377 + 2852 =

Estimate:

f   8348 + 4271 =

Estimate:

Number

3 Grace has used the formal written method to answer three addition calculations. Identify any mistakes that she has made and correct them, using squared paper to show your working.

Then write what you would say to Grace to help her avoid making these mistakes in future.

a

		8	0	8	9
+		9	3	5	4
	1	0	1	6	2	9
		1			1	

b

		5	6	8	5
+		8	2	7	3
	1	4	8	5	8
		1			

c

	4	3	7	6
+	2	8	3	4
	6	1	0	0
		1	1	1

_____

_____

_____

4 Which of these calculations gives a sum that reads the same left to right as it does right to left, for example, 34 743? ☐

Use squared paper to show your working.

a 2376 + 1858 =

b 4689 + 2258 =

c 8653 + 3389 =

d 5196 + 7725 =

5 The number of people visiting four museums (**a–d**) is recorded in the morning and the afternoon on one day.

Which museum had the greatest number of visitors?

Estimate: ☐   Answer: ☐

Use squared paper to show your working.

a

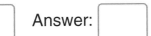

MUSEUM
Morning
42 435
Afternoon
28 283

b

MUSEUM
Morning
36 284
Afternoon
34 617

c

MUSEUM
Morning
19 064
Afternoon
51 638

d

MUSEUM
Morning
45 829
Afternoon
25 771

Date: _____

Lesson 3: Adding positive and negative numbers

- Use a number line to add a positive number to a negative number

1 Use the number track to find each sum. Begin at the first addend and count on to the right the number of places given by the second addend.

−10	−9	−8	−7	−6	−5	−4	−3	−2	−1	0	1	2	3	4	5	6	7	8	9	10

a $-7 + 2 = \boxed{}$ **b** $-5 + 1 = \boxed{}$ **c** $-9 + 3 = \boxed{}$

d $-6 + 2 = \boxed{}$ **e** $-8 + 4 = \boxed{}$ **f** $-9 + 5 = \boxed{}$

g $-7 + 4 = \boxed{}$ **h** $-6 + 5 = \boxed{}$ **i** $-2 + 2 = \boxed{}$

j $-4 + 4 = \boxed{}$ **k** $-6 + 6 = \boxed{}$ **l** $-10 + 10 = \boxed{}$

m $-3 + 4 = \boxed{}$ **n** $-2 + 5 = \boxed{}$ **o** $-4 + 6 = \boxed{}$

p $-8 + 9 = \boxed{}$ **q** $-5 + 10 = \boxed{}$ **r** $-7 + 9 = \boxed{}$

2 Use the number line to find each answer.

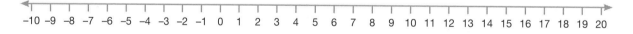

a $-2 + 5 = \boxed{}$ **b** $-9 + 7 = \boxed{}$ **c** $-5 + 5 = \boxed{}$

d $-8 + 9 = \boxed{}$ **e** $-4 + 3 = \boxed{}$ **f** $-2 + 9 = \boxed{}$

g $-7 + 6 = \boxed{}$ **h** $-9 + 9 = \boxed{}$ **i** $-4 + 12 = \boxed{}$

j $-6 + 13 = \boxed{}$ **k** $-5 + 15 = \boxed{}$ **l** $-8 + 14 = \boxed{}$

m $-7 + 15 = \boxed{}$ **n** $-1 + 18 = \boxed{}$ **o** $-2 + 16 = \boxed{}$

3 Calculate the new temperatures. Write each calculation as a number sentence.

Start temperature	−4°C	−6°C	−2°C	−8°C	−1°C	−9°C
Rise	9 degrees	12 degrees	15 degrees	16 degrees	19 degrees	20 degrees
Calculation	−4 + 9 =					
New temperature						

4 Nine people put money into their bank accounts, which all have a negative amount.
Find the new amount of money in their accounts and write the letter codes of the accounts in order, from least to greatest.

A	−$10 + $13		B	−$8 + $10		C	−$5 + $15

D	−$9 + $14		E	−$7 + $13		F	−$12 + $19

G	−$11 + $15		H	−$4 + $12		I	−$6 + $15

Order:

☐ < ☐ < ☐ < ☐ < ☐ < ☐ < ☐ < ☐ < ☐

5 Write two numbers, one negative and one positive, that will give each sum.

a ☐ + ☐ = 0   **b** ☐ + ☐ = −2   **c** ☐ + ☐ = 2

d ☐ + ☐ = −1   **e** ☐ + ☐ = 5   **f** ☐ + ☐ = −3

g ☐ + ☐ = 7   **h** ☐ + ☐ = −6

6 Fuhua solves −3 + 18 by starting at −3 and counting on 18 steps of 1 until she reaches 15. Could Fuhua have used a better, more efficient method to add 15? If so, describe it.

Date: _____

😊 😐 ☹

Number

Lesson 4: **Identifying values for symbols in calculations**

• Find the value of unknown quantities represented by symbols in calculations

1 Work out the unknown values.

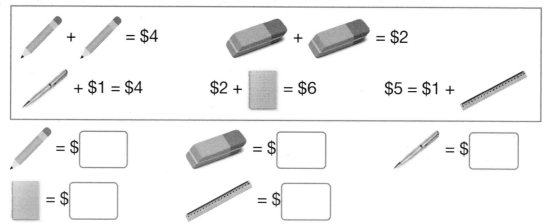

= $ [ ]   = $ [ ]   = $ [ ]

= $ [ ]   = $ [ ]

2 Work out the unknown values.

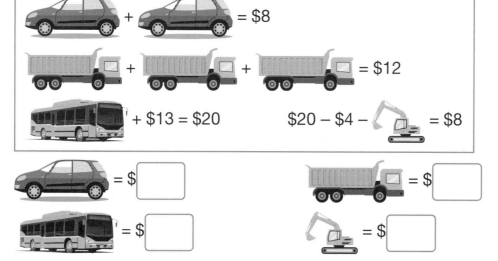

= $ [ ]   = $ [ ]

= $ [ ]   = $ [ ]

3 Use the number sentences to work out the price of each item.

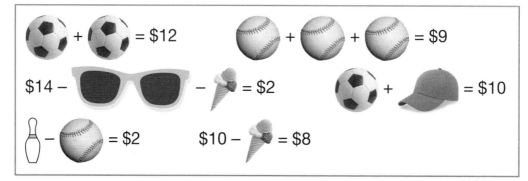

 = $ ☐

= $ ☐

= $ ☐

= $ ☐

= $ ☐

= $ ☐

4 Find the value of each symbol.

a $5 + \square = 13$

$\square =$ ☐

b $\bigcirc - 2 = 8$

$\bigcirc =$ ☐

c $\triangle + \triangle + \triangle = 27$

$\triangle =$ ☐

d $8 + \left(+ * = 24 \right.$

$* + * + 2 = 22$

$* =$ ☐ , $\left(= \right.$ ☐

e $20 - \diamond - \maltese = 7$

$15 - \maltese = 8$

$\maltese =$ ☐ , $\diamond =$ ☐

5 Write number sentences for each word problem, using symbols to represent unknown values. Work out the value of each symbol and the cost of each item.

a Taran buys four ice creams for $12. Leah buys an ice cream and a chocolate bar for $5. What is the price of each item?

ice cream: $ ☐   chocolate bar: $ ☐

b Ria uses a $10 note to pay for a magazine and a lollipop and receives $4 change. Jamie uses a $5 note to pay for a lollipop and receives $4 in change. What is the price of each item?

magazine: $ ☐   lollipop: $ ☐

Date: _____

21

Lesson 1: **Subtracting 4-digit numbers (1)**

- Estimate and subtract 4-digit numbers, choosing the best mental or written method

You will need
- Resource sheet 3: 2 cm squared paper
- ruler

1 Solve each calculation.

a 443 – 199 = ☐

b 473 – 251 = ☐

c 870 – 480 = ☐

d 923 – 298 = ☐

e 875 – 225 = ☐

f 873 – 586 = ☐

2 Use the expanded written method.

a 8376 – 2844 = ☐

	8000	300	70	6
–	2000	800	40	4

☐

b 7466 – 3287 = ☐

	7000	400	60	6
–	3000	200	80	7

☐

c 4541 – 1775 = ☐

	4000	500	40	1
–	1000	700	70	5

☐

d 5342 – 2567 = ☐

	5000	300	40	2
–	2000	500	60	7

☐

Number

3 Use the expanded written method to answer these calculations. Estimate the answer first. Use squared paper if you need to.

a 3634 − 1462 = ☐

Estimate: ☐

b 4573 − 2246 = ☐

Estimate: ☐

c 7473 − 2741 = ☐

Estimate: ☐

d 5684 − 2736 = ☐

Estimate: ☐

e 9458 − 4775 = ☐

Estimate: ☐

f 6944 − 3587 = ☐

Estimate: ☐

4 Rope A measures 5673 cm in length and Rope B measures 7242 cm in length. 2857 cm is cut from Rope A and 4466 cm is cut from Rope B. Which rope is now longer? Show your working.

Rope A (cut length): ☐ cm   Rope B (cut length): ☐ cm

Longer rope: ☐

☐

5 Use the expanded written method to answer these calculations. Estimate the answer first. Use 2 cm squared paper if you need to.

a 42734 − 25223 = ☐

Estimate: ☐

b 56387 − 33855 = ☐

Estimate: ☐

c 68337 − 55182 = ☐

Estimate: ☐

d 25428 − 18276 = ☐

Estimate: ☐

Date: _____

Lesson 2: **Subtracting 4-digit numbers (2)**

- Estimate and subtract 4-digit numbers using the formal written method

You will need
- squared paper
- ruler

1 Complete each calculation.

a
```
    6 5 7
-   2 3 4
```

b
```
    8 6 8
-   6 4 5
```

c
```
    7 7 9
-   5 6 9
```

d
```
  7 6 8 5
- 4 3 4 8
```

e
```
  8 8 3 6
- 6 6 5 3
```

f
```
  9 2 8 8
- 5 8 7 7
```

2 Use the formal written method to calculate the difference. Estimate the answer first. Use squared paper to show your working.

a 7844 – 3372 =

Estimate:

b 6853 – 2921 =

Estimate:

c 8493 – 5763 =

Estimate:

d 9132 – 6515 =

Estimate:

e 5293 – 2747 =

Estimate:

f 3478 – 1883 =

Estimate:

g 7432 – 4666 =

Estimate:

h 8437 – 3668 =

Estimate:

3 Alex has answered three subtraction calculations using the formal written method. Identify any mistakes that he has made and correct them using squared paper to show your working.

Then write how you would convince Alex of a method that would help him avoid making these mistakes in future.

a

	3	7	4	¹3
−	1	4	2	8
	1	3	2	5

b

	⁷8̸	¹2̸	¹6̸	7
−	4	7	8	2
	3	5	8	5

c

	7	¹⁶6̸⁷	¹³7̸	4̸	¹6̸
−	3	8	6	8	
	4	8	3	8	

_____

_____

_____

4 Order the answers to these calculations, from smallest to greatest.

$4876 - 1488 =$ $7557 - 4175 =$ $7844 - 4556 =$

$6432 - 2977 =$ $9035 - 5658 =$

[ ] < [ ] < [ ] < [ ] < [ ]

5 Write in the missing numbers.

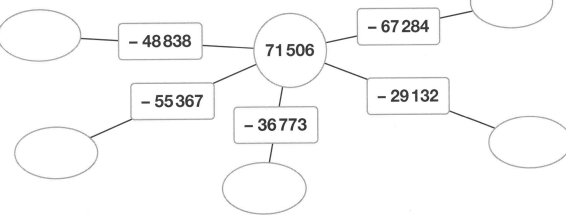

Date: _____

Lesson 3: **Subtraction where the answer is a negative number**

- Subtracting pairs of numbers where the answer is a negative number

1 Use the number track to solve each calculation. Begin at the minuend and count back the number of places given by the subtrahend.

-10	-9	-8	-7	-6	-5	-4	-3	-2	-1	0	1	2	3	4	5	6	7	8	9	10

a $-1 - 1 =$ ☐

b $-2 - 1 =$ ☐

c $-3 - 1 =$ ☐

d $-4 - 1 =$ ☐

e $-1 - 2 =$ ☐

f $-2 - 2 =$ ☐

g $-2 - 3 =$ ☐

h $-2 - 4 =$ ☐

i $1 - 2 =$ ☐

j $1 - 3 =$ ☐

k $1 - 4 =$ ☐

l $1 - 5 =$ ☐

m $2 - 3 =$ ☐

n $2 - 4 =$ ☐

o $2 - 5 =$ ☐

p $2 - 6 =$ ☐

q $4 - 5 =$ ☐

r $4 - 6 =$ ☐

 2 Use the number line to find each answer.

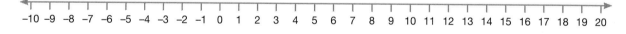
-10 -9 -8 -7 -6 -5 -4 -3 -2 -1 0 1 2 3 4 5 6 7 8 9 10 11 12 13 14 15 16 17 18 19 20

a $3 - 5 =$ ☐

b $6 - 8 =$ ☐

c $-3 - 6 =$ ☐

d $-4 - 5 =$ ☐

e $5 - 7 =$ ☐

f $-4 - 3 =$ ☐

g $3 - 8 =$ ☐

h $5 - 9 =$ ☐

i $-1 - 7 =$ ☐

j $-6 - 2 =$ ☐

k $7 - 10 =$ ☐

l $-9 - 1 =$ ☐

m $0 - 9 =$ ☐

n $2 - 10 =$ ☐

o $-7 - 3 =$ ☐

3 Calculate the new temperatures. Write each calculation as a number sentence.

Start temperature	−3 °C	5 °C	−6 °C	7 °C	−9 °C	11 °C
Decrease	9 degrees	11 degrees	7 degrees	14 degrees	15 degrees	18 degrees
Calculation	−3 − 9 =					
New temperature						

4 A bank contacts a customer if their account falls below −$16.
Nine customers all made recent purchases. Write the letter codes of the account holders the bank will need to contact.
Find the new amount of money in their accounts and write the letter codes of the accounts in order, from least to greatest.

A	−$4 − $11
B	$3 − $18
C	−$5 − $12

D	−$8 − $7
E	$5 − $22
F	−$7 − $9

G	$13 − $30
H	−$1 − $16
I	$23 − $40

The bank needs to contact account holders: _____

Order:

☐ < ☐ < ☐ < ☐ < ☐ < ☐ < ☐ < ☐ < ☐

5 For each calculation, write two positive numbers that will give the answer.

a ☐ − ☐ = −3   **b** ☐ − ☐ = −1   **c** ☐ − ☐ = −6

For each calculation, write two negative numbers that will give the answer.

d ☐ − ☐ = −9   **e** ☐ − ☐ = −14   **f** ☐ − ☐ = −21

Date: _____

Number

Lesson 4: **Identifying values for symbols in subtraction calculations**

- Find the value of unknown quantities in subtraction calculations that are represented by symbols

1 Work out the unknown values.

$1 = $2 − $3 = $5 − $4 = $7 −

− $1 = $2   − $2 = $3

= $ ☐   = $ ☐   = $ ☐

= $ ☐   = $ ☐

2 Work out the unknown values.

− $4 = $9   − $6 = $11

− $8 = $13 $5 = $23 − $9 = $41 −

= $ ☐   = $ ☐

= $ ☐   = $ ☐   = $ ☐

3 Use the number sentences to work out the price of each item.

$30 − = $5 $7 = $80 − −

$50 − − = $5 $11 = $40 −

 = $ ☐    = $ ☐    = $ ☐    = $ ☐

4 Find the value of each symbol.

  a $11 - \square = 3$

    $\square = \boxed{}$

  b $\bigcirc - 14 = 33$

    $\bigcirc = \boxed{}$

  c $53 - ☾ - ✳ = 18$

    $32 - ✳ = 17$

    $✳ = \boxed{}$ $☾ = \boxed{}$

Number

5 Write number sentences for each word problem using symbols to represent unknown values. Work out the value of each symbol and the cost of each item.

  a Jake buys a rucksack. He pays $50 and receives $16 in change. Fatimah buys a rucksack and a water bottle. She pays $100 and receives $49 in change.

  Rucksack: $\boxed{}$ Water bottle: $\boxed{}$

  b Freya buys a laptop bag. She pays $40 and receives $7 in change. Jayesh buys a laptop bag and a video game. He pays $100 and receives $29 in change.

  Laptop bag: $\boxed{}$ Video game: $\boxed{}$

Date: _____

Number

Lesson 1: **Prime and composite numbers (1)**

- Understand and explain the difference between prime and composite numbers

You will need
- coloured pencils

1 Find all the factors of each number. Use the factor rainbows to help you.

a 7: [   ]   **b** 21: [   ]   **c** 17: [   ]

d 32: [   ]   **e** 18: [   ]   **f** 70: [   ]

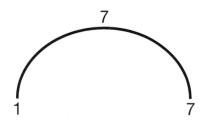

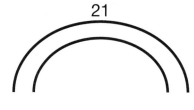

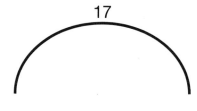

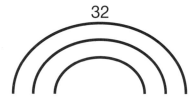

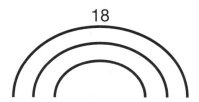

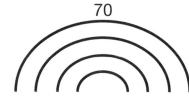

2 Draw your own factor rainbows to find all the factors for each number.

a 80

b 72

Number

3 **a** Is the number 17 a prime number? Explain your answer.

_____

b Is the number 34 a prime number? Explain your answer.

_____

c Is the number 1 a prime number? Explain your answer.

_____

d Is the number 2 a prime number? Explain your answer.

_____

4 Colour the prime numbers in each set.

a

| 50 | 51 | 52 | 53 | 54 | 55 | 56 | 57 | 58 | 59 | 60 |

b

| 20 | 21 | 22 | 23 | 24 | 25 | 26 | 27 | 28 | 29 | 30 |

c

| 80 | 81 | 82 | 83 | 84 | 85 | 86 | 87 | 88 | 89 | 90 |

d

| 30 | 31 | 32 | 33 | 34 | 35 | 36 | 37 | 38 | 39 | 40 |

Which set has the most prime numbers? [   ]

5 **a** Which two consecutive prime numbers below 20 have the:

   i smallest difference [   ]   **ii** greatest difference? [   ]

b Which two consecutive prime numbers between 20 and 40 have the:

   i smallest difference [   ]   **ii** greatest difference? [   ]

Date: _____

Number

Lesson 2: **Prime and composite numbers (2)**

- Understand and explain the difference between prime and composite numbers

You will need

- blue and red coloured pencils

 1 Draw a ring around the prime numbers red and the composite numbers blue.

	2	3	4	5
6	7	8	9	10
11	12	13	14	15
16	17	18	19	20

Eratosthenes

 2 Use the sieve of Eratosthenes method to 'drain' out all the composite numbers and leave all the prime numbers up to 100 behind.

List the prime numbers here.

	2	3	4	5	6	7	8	9	10
11	12	13	14	15	16	17	18	19	20
21	22	23	24	25	26	27	28	29	30
31	32	33	34	35	36	37	38	39	40
41	42	43	44	45	46	47	48	49	50
51	52	53	54	55	56	57	58	59	60
61	62	63	64	65	66	67	68	69	70
71	72	73	74	75	76	77	78	79	80
81	82	83	84	85	86	87	88	89	90
91	92	93	94	95	96	97	98	99	100

3 Complete the calculations and draw a ring around all the answers that are prime.

a   32 + 49 = ☐   **b**   6 × 13 = ☐   **c**   97 − 58 = ☐

d   72 ÷ 4 = ☐   **e**   28 + 27 = ☐   **f**   3 × 26 = ☐

g   114 − 35 = ☐   **h**   63 ÷ 9 = ☐   **i**   39 + 58 = ☐

j   14 × 7 = ☐   **k**   103 − 56 = ☐   **l**   87 ÷ 3 = ☐

Do you have to complete each calculation to know whether the answer is prime? Explain your reasoning.

_____

_____

4 George says:

> If you multiply two prime numbers, other than the number 2, you always get an odd number.

Is George correct? Convince George of why you think that.

_____

_____

5 Arrange the digits 1 to 9 in the squares of a 3 by 3 grid so that the numbers in every row and every column add up to a prime number. How many ways of arranging the numbers can you find? One has been done for you.

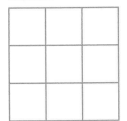

2	8	1
4	6	7
5	9	3

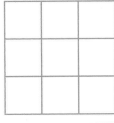

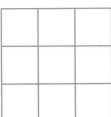
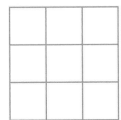

Date: _____

Lesson 3: **Tests of divisibility: 4 and 8**

• Identify numbers that are divisible by 4 and 8

1 Write the missing numbers.

a $2 \times 8 =$ ☐   **b** $6 \times 4 =$ ☐   **c** $3 \times 8 =$ ☐

d $9 \times 2 =$ ☐   **e** $7 \times 3 =$ ☐   **f** $11 \times 2 =$ ☐

g $4 \times 9 =$ ☐   **h** $8 \times 7 =$ ☐   **i** $6 \times$ ☐ $= 12$

j ☐ $\times 8 = 16$ **k** $4 \times$ ☐ $= 20$ **l** ☐ $\times 5 = 40$

m $12 \times$ ☐ $= 24$ **n** ☐ $\times 4 = 24$ **o** $2 \times$ ☐ $= 18$

2 Draw a ring around the numbers that are divisible by 2.

422   470   243   335   876   918   777

1231   2440   4784   3849   7765   8088   9122

 3 a Draw a ring around the numbers that are divisible by 4.

316   4788   5356   2631

477   8968   3644   48 394

b Draw a ring around the numbers that are divisible by 8.

9599   8512   8151   57 472

5917   1164   4354   30 448

4

Use the numbers on the cards to fill in the gaps. Then complete each sentence by writing the related divisibility rule. Each number can be used only once. The first one has been classified for you.

| 7984 | 634 | 3876 | 4 | 8 | 472 | 5512 | 8 |

a ┌ 634 ┐ is divisible by 2 because <u>*it ends in an even number.*</u>

b ┌___┐ is divisible by 4 because _____

c 2432 is divisible by ┌___┐ because _____

d ┌___┐ is divisible by 8 because _____

e 2716 is divisible by ┌___┐ because _____

f ┌___┐ is divisible by 4 because _____

g ┌___┐ is divisible by ┌___┐ because _____

5 Use the digits 1 to 9 to complete the numbers.

Divisible by 4:

a 8 3 ┌___┐

b 6 5 ┌___┐

c 7 7 ┌___┐

d 4 9 1 ┌___┐

e 7 8 8 ┌___┐

f 9 0 9 ┌___┐

g 6 2 ┌___┐┌___┐

h 1 0 ┌___┐┌___┐

i 8 4 ┌___┐┌___┐

Divisible by 8:

j 7 0 3 ┌___┐

k 5 1 6 ┌___┐

l 3 2 4 ┌___┐

m 1 2 6 ┌___┐

n 2 3 4 ┌___┐

o 6 1 7 ┌___┐

p 6 2 ┌___┐ 2

q 4 4 ┌___┐ 8

r 8 3 ┌___┐ 8

Number

Date: _____

35

Lesson 4: **Square numbers**

- Recognise square numbers from 1 to 100

1 Complete the multiplications.

a $3 \times 3 = $ ⬚

b $1 \times 1 = $ ⬚

c $5 \times 5 = $ ⬚

d $2 \times 2 = $ ⬚

e $10 \times 10 = $ ⬚

f $4 \times 4 = $ ⬚

g $9 \times 9 = $ ⬚

h $7 \times 7 = $ ⬚

i $8 \times 8 = $ ⬚

2 Solve the problems.

a Dara has 6 bags of shells. If each bag contains 6 shells, how many shells does he have in total? ⬚

b Mia has 9 packs of stickers. If each pack has 9 stickers, how many stickers does she have in total? ⬚

3 Complete the statements.

Example: The square of 1 or 1^2 is 1. 1 is a square number.

a The square of 4 or ⬚ ⬚ is ⬚ . ⬚ is a square number.

b The square of 7 or ⬚ ⬚ is ⬚ . ⬚ is a square number.

c The square of 9 or ⬚ ⬚ is ⬚ . ⬚ is a square number.

d The square of 3 or ⬚ ⬚ is ⬚ . ⬚ is a square number.

e The square of 8 or ⬚ ⬚ is ⬚ . ⬚ is a square number.

4 Write the missing numbers in the sequence of squares.

4 ⬚ 16 ⬚ 36 ⬚ 64 ⬚ 100

Number

Number

5 Draw a ring around the square numbers in each group.

a | 44 48 19 4 98 64 21 | **b** | 58 3 54 9 85 25 46

c | 76 5 16 81 62 23 36 | **d** | 49 10 23 15 97 100 6

6 Find the numbers.

a The sum of two consecutive square numbers is 41.

The two numbers that are squared are [ ] and [ ].

b The sum of two consecutive square numbers is 145.

The two numbers that are squared are [ ] and [ ].

c The difference between two consecutive square numbers is 19.

The two numbers that are squared are [ ] and [ ].

7 **a** Find two square numbers that add together to make another square number.

[ ] + [ ] = [ ]

b Find three square numbers that add together to make another square number.

[ ] + [ ] + [ ] = [ ]

Date: _____

Lesson 1: **Simplifying calculations (1)**

* Know which property of number to use to simplify calculations

1 Write the missing numbers.

a $34 \times 6 = \boxed{} \times 3$

b $46 \times 17 \times 5 = 5 \times \boxed{} \times 17$

c $7 \times 43 = \boxed{7 \times \boxed{}} + \boxed{7 \times 3}$

d $8 \times 74 = \boxed{} \times 8$

e $5 \times 24 \times 8 = 8 \times \boxed{} \times 24$

f $96 \times 8 = \boxed{90 \times \boxed{}} + \boxed{\boxed{} \times 8}$

2 Solve each calculation mentally. Question **a** and part of question **b** have been done for you.

a $2 \times 8 \times 5 = \boxed{2 \times 5 \times 8} = \boxed{10 \times 8} = \boxed{80}$

b $6 \times 53 = \boxed{6 \times 50} + \boxed{6 \times 3} = \boxed{300} + \boxed{18} = \boxed{}$

c $4 \times 74 = \boxed{} + \boxed{} = \boxed{} + \boxed{} = \boxed{}$

d $5 \times 9 \times 4 = \boxed{} = \boxed{} = \boxed{}$

e $7 \times 67 = \boxed{} + \boxed{} = \boxed{} + \boxed{} = \boxed{}$

f $6 \times 13 \times 5 = \boxed{} = \boxed{} = \boxed{}$

g $8 \times 93 = \boxed{} + \boxed{} = \boxed{} + \boxed{} = \boxed{}$

h $5 \times 18 \times 8 = \boxed{} = \boxed{} = \boxed{}$

3 Work out the answer to each calculation. Then rewrite the calculation using the associative property and find the answer. The first one has been done for you.

a $4 \times 6 \times 2 = \boxed{24 \times 2} = \boxed{48}$ $4 \times 6 \times 2 = \boxed{4 \times 12} = \boxed{48}$

b $8 \times 4 \times 3 = \boxed{} = \boxed{}$ $8 \times 4 \times 3 = \boxed{} = \boxed{}$

c $9 \times 7 \times 2 = \boxed{} = \boxed{}$ $9 \times 7 \times 2 = \boxed{} = \boxed{}$

d $12 \times 3 \times 5 = \boxed{} = \boxed{}$ $12 \times 3 \times 5 = \boxed{} = \boxed{}$

Number

Number

4 Calculate the answer to each problem. Write your working and state the number property you have used to simplify the calculation.

a 5 tables are prepared for a party. 4 trays are placed on each table and 23 cakes placed on each tray. How many cakes are there altogether?

Property: _____

b There are 27 red bags and 45 green bags. Each bag holds 2 footballs. How many footballs are there altogether?

Property: _____

c There are 68 rooms on each floor of a hotel. The hotel has 7 floors. How many rooms are there in total?

Property: _____

5 Solve this multi-step problem.

6 buildings are cube-shaped and identical in design. Each building has 6 floors. There are 5 windows on each side of the building per floor. How many windows will all 6 buildings have in total?

Clue: Don't forget that each floor has 4 sides!

Date: _____

Lesson 2: **Simplifying calculations (2)**

• Know which property of number to use to simplify calculations

1 Use the distributive property to complete each calculation. Question **a** and part of questions **b** and **c** have been done for you.

a 37 × 4 = | 30 × 4 | + | 7 × 4 |

= | 120 | + | 28 |

= | 148 |

b 83 × 3 = | | + | |

= | 240 | + | 9 |

= | 249 |

c 48 × 6 = | 40 × 6 | + | 8 × 6 |

= | | + | |

= | |

d 54 × 8 = | | + | |

= | | + | |

= | |

e 66 × 7 = | | + | |

= | | + | |

= | |

f 83 × 8 = | | + | |

= | | + | |

= | |

2 Use the distributive property to find each product.

a 12 × 9 =

b 9 × 15 =

c 4 × 58 =

d 47 × 8 =

e 23 × 7 =

f 67 × 6 =

Number

3 Use diagrams and examples to explain the distributive property of multiplication.

4 The scale is balanced. Can you use the distributive property of multiplication to convincingly explain this?

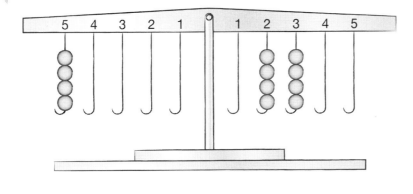

_____

_____

5 Write four different ways of using the distributive property to multiply 8 × 376.

Date: _____

Number

Lesson 3: **Order of operations (1)**

• Understand that the four operations of number follow a particular order

1 Use the order of operations to calculate. Show any working.

a $2 \times 6 + 3 =$ ⬚

b $5 + 2 - 1 =$ ⬚

c $6 \div 2 + 11 =$ ⬚

d $12 - 5 \times 2 =$ ⬚

e $9 + 9 - 7 =$ ⬚

f $4 \times 5 + 6 =$ ⬚

g $10 \div 2 - 5 =$ ⬚

h $10 - 8 \times 3 =$ ⬚

i $3 + 15 - 4 =$ ⬚

2 Use the order of operations to calculate. Show any working.

a $8 + 3 \times 2 =$ ⬚

b $42 \div 7 - 4 =$ ⬚

c $6 \times 5 + 7 =$ ⬚

d $14 + 45 \div 5 =$ ⬚

e $9 \div 3 + 7 =$ ⬚

f $47 - 27 \div 3 =$ ⬚

g $16 - 10 \div 2 =$ ⬚

h $99 - 72 \div 9 =$ ⬚

i $24 + 49 \div 7 =$ ⬚

3 Mia has answered a set of calculations. Identify and explain any mistakes she has made and correct them.

4

a $12 + 4 \times 3 = 23$

b $72 \div 9 - 1 = 9$

c $12 \times 5 + 2 = 84$

d $36 - 12 \div 3 = 32$

e $23 + 9 \times 7 = 224$

f $48 \div 8 - 6 = 24$

4 Using each of the numbers 1, 2 and 4 only once, and two operators (add or subtract and multiply or divide), how many different calculations can you write and solve?

1

Date: _____

Number

Lesson 4: **Order of operations (2)**

• Understand that the four operations of number follow a particular order

1 Use the order of operations to calculate.

a $7 \times 2 + 4 =$ ☐   **b** $6 - 4 \div 2 =$ ☐   **c** $27 \div 3 + 6 =$ ☐

d $15 + 13 \times 3 =$ ☐   **e** $16 \times 3 - 1 =$ ☐   **f** $30 \div 6 - 5 =$ ☐

g $42 + 36 \div 6 =$ ☐   **h** $31 - 3 \times 9 =$ ☐   **i** $45 + 15 \times 4 =$ ☐

2 Write a calculation to represent each word problem and then solve it. Remember to use the order of operations to calculate.

a Max bought 4 books for $9 each and 1 book for $12.

How much did Max pay in total? ☐

b Cai bought 3 necklaces for $25 each and paid with a $100 note.

How much change does Cai receive from the shop assistant? ☐

c Mr Khan mends bikes. He charges $15 an hour for his work. It took 3 hours to mend Elif's bike. He also used parts costing $48.

What will he charge Elif? ☐

Number

d Fred has 90 football stickers. He takes 60 of the stickers and divides them into 5 equal groups. He gives one of these groups to a friend.

How many stickers does he have left?

3 Pavel has answered a set of calculations. Identify and explain any mistakes he has made and correct them.

a $5 + 6 \times 2 = 22$

b $48 \div 8 - 3 = 3$

c $9 + 4 \times 4 = 25$

d $9 + 18 \div 3 = 9$

4 Find the missing numbers or operations ($+$, $-$, $\times$ or $\div$). Show any working out.

a $26 + \boxed{} \times 8 = 42$

b $30 - \boxed{} \div 4 = 27$

c $56 - \boxed{} \times 11 = 12$

d $8 \boxed{} 3 \boxed{} 4 = 20$

Date: _____

Number

Lesson 1: **Working with place value**

• Use place value to multiply numbers to 1000 by 1-digit numbers

1 Decompose each of these numbers.

Example:

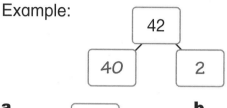

a
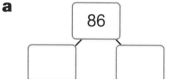

b
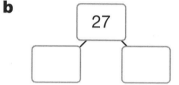

c
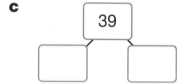

d
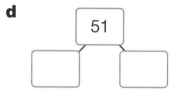

e
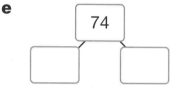

f
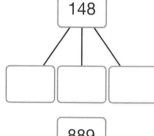

g
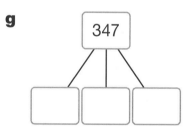

h
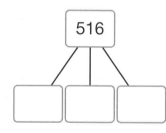

i
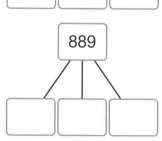

2 Use the diagram to answer the questions.

a Ryan has arranged place value counters to show a calculation.

(100)(100)(100)(100)	(10)(10)	(1)(1)(1)(1)(1)(1)
(100)(100)(100)(100)	(10)(10)	(1)(1)(1)(1)(1)(1)
(100)(100)(100)(100)	(10)(10)	(1)(1)(1)(1)(1)(1)
(100)(100)(100)(100)	(10)(10)	(1)(1)(1)(1)(1)(1)

What is the value of the counters in each row?

100s: ☐   10s: ☐   1s: ☐

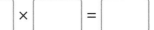

b How many rows are there?

100s: ☐   10s: ☐   1s: ☐

c Write and solve the calculation represented by each group of counters.

100s: ☐ × ☐ = ☐   10s: ☐ × ☐ = ☐

1s: ☐ × ☐ = ☐

d What calculation is modelled by the counters? ☐ × ☐ = ☐

 Draw the arrangement of place value counters to show the calculation and then use it to find the product. Estimate your answer first.

a 7 × 43 = ☐

Estimate: ☐

10s	1s

b 6 × 342 = ☐

Estimate: ☐

100s	10s	1s

4 Draw lines to match each calculation with its product.

4 × 83		47 × 7		8 × 294

	3 × 786		263 × 9		54 × 6

329		332		324

	2352		2367		2358

Date: _____

Number

Lesson 2: **Grid method**

Number

- Use the grid method to multiply numbers to 1000 by 1-digit numbers

1 Partition the numbers in the grid. Then work out the answer.

a 18 × 4 = ☐

× ☐ ☐

☐ ☐ ☐

☐

b 24 × 5 = ☐

× ☐ ☐

☐ ☐ ☐

☐

c 17 × 8 = ☐

× ☐ ☐

☐ ☐ ☐

☐

d 21 × 6 = ☐

× ☐ ☐

☐ ☐ ☐

☐

2 For each calculation, estimate first, then use the grid to work out the answer. Show your working. Check your answer with your estimate.

a 197 × 3 = ☐

Estimate: ☐

× ☐ ☐ ☐

☐ ☐ ☐ ☐

☐

b 839 × 8 = ☐

Estimate: ☐

× ☐ ☐ ☐

☐ ☐ ☐ ☐

☐

Number

3 For each calculation, estimate first, then use the grid method to work out the answer. Show your working. Check your answer with your estimate.

a 437 × 2 =

Estimate:

b 698 × 4 =

Estimate:

c 808 × 6 =

Estimate:

4 Use these number cards to write two different word problems where a 3-digit number is multiplied by a 1-digit number.

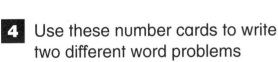 473 7 264 4 786 8

a _____

_____

_____

b _____

_____

_____

Use the grid method to solve your word problems. Estimate first.

Date: _____

Lesson 3: **Expanded written method**

- Use the expanded written method to multiply numbers to 1000 by 1-digit numbers

1 Round and estimate first, then multiply.

a Estimate:

```
        3  6
  ×        4
  ─────────────
             ×
  +          ×
  ─────────────
```

b Estimate:

```
        7  8
  ×        5
  ─────────────
             ×
  +          ×
  ─────────────
```

2 Round and estimate first, then multiply.

a Estimate:

```
     5  1  6
  ×        4
  ─────────────
             ×
             ×
  +          ×
  ─────────────
```

b Estimate:

```
     4  8  3
  ×        6
  ─────────────
             ×
             ×
  +          ×
  ─────────────
```

c Estimate:

```
     7  2  7
  ×        7
  ─────────────
             ×
             ×
  +          ×
  ─────────────
```

d Estimate:

```
     9  8  7
  ×        8
  ─────────────
             ×
             ×
  +          ×
  ─────────────
```

3 Write in the missing numbers.

a

			3
×			3
			2
			0
+	1	5	0	0
	1		0	

4	×	3
	×	3
	×	3

b

		6		2
×				
				8
		2	8	0
+			0	0
			6	8

	×	4
	×	
	×	4

c

			8
×			
		6	4
	4	0	0
+	7	2	0	0

	×	8
	×	
	×	

d

		5	
×			
	5	6	
	4	9	0
+			

	×	
	×	7
500	×	

4 Order the baskets by mass, from lightest to heaviest.

Basket A

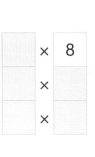

286 g 286 g 286 g 286 g 286 g

Basket B

167 g 167 g 167 g 167 g 167 g 167 g 167 g 167 g 167 g

Basket C

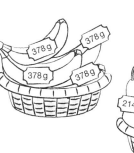

378 g 378 g 378 g 378 g

Basket D

214 g 214 g 214 g 214 g 214 g 214 g 214 g

Basket E

478 g 478 g 478 g 478 g

Order: [   ] , [   ] , [   ] , [   ] , [   ]

Date: _____

Number

Lesson 4: **Real-life problems**

- Solve problems involving multiplication of numbers to 1000 by 1-digit numbers

You will need
- paper
- coloured pencils

1 Draw a ring around the greater amount.

a

b

2 Solve the problems. Use paper to show your working. Remember to estimate first.

a A recipe uses 7 measures of 127 g of sugar.

How much sugar does the recipe require? ☐ g

b A museum is allowed 433 visitors per day. How many visitors are allowed into the museum in 6 days?

☐ visitors

c Jugs contain 778 ml of orange juice.

What is the total volume of juice in 5 jugs? ☐ ml

d Toby scored 946 points in each of 7 video games.

How many points did he score in total? ☐ points

Number

 3 Ria plays a dartboard game. She must only hit the pairs of sectors (one inner and one outer) that multiply to give a product between 3400 and 3600. Help her by colouring the sectors where she should aim a dart. Use paper to show your working.

One sector has been completed for you.

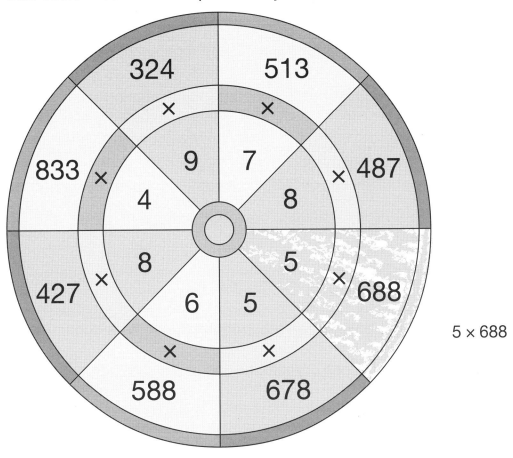

5 × 688

4 A shop sells sandwiches. In the first week, they sold 237 sandwiches a day. In the second week they sold 376 sandwiches a day. In the third week, they sold 456 sandwiches a day. How many sandwiches did they sell altogether. Estimate the answer first.

Date: _____

Number

Lesson 1: **Working with place value**

- Use place value to multiply numbers to 1000 by 2-digit numbers

You will need
- place value counters
- area model chart

1 Aarif has arranged place value counters to show a calculation.

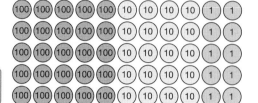

a What is the value of the counters in each row?

100s: ▢   10s: ▢   1s: ▢

b How many rows are there?

100s: ▢   10s: ▢   1s: ▢

c Write and solve the calculation represented by each group of counters.

100s: ▢ × ▢ = ▢   10s: ▢ × ▢ = ▢

1s: ▢ × ▢ = ▢

d What calculation is modelled by the counters?   ▢ × ▢ = ▢

2 Use the distributive property of multiplication to simplify the calculations. The first one has been done for you. You do not have to work out the answer.

$23 \times 27 =$ | $20 + 3$ | × | $20 + 7$

= | 20×20 | + | 20×7 | + | 3×20 | + | 3×7

a $32 \times 44 =$ ▢ × ▢

= ▢ + ▢ + ▢ + ▢

b $46 \times 58 =$ ▢ × ▢

= ▢ + ▢ + ▢ + ▢

c $38 \times 63 =$ ▢ × ▢

= ▢ + ▢ + ▢ + ▢

Model each multiplication using place value counters on an area model chart and then use it to find the product.

a 17 × 23 = ⬜   **b** 21 × 26 = ⬜   **c** 27 × 34 = ⬜

d 24 × 33 = ⬜   **e** 26 × 37 = ⬜   **f** 28 × 38 = ⬜

 Use the area diagram to calculate the product. The first one has been done for you.

a 37 × 46 = ⬜

×	40	6
30	30 × 40 = 1200	30 × 6 = 180
7	7 × 40 = 280	7 × 6 = 42

1200 + 180 + 280 + 42 = 1702

b 33 × 38 = ⬜

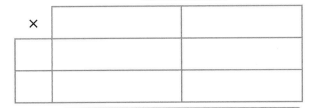

c 44 × 47 = ⬜

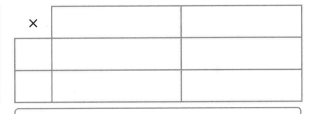

d 47 × 53 = ⬜

5 Sara has used place value counters to model four 2-digit by 2-digit multiplications. Each addition below is the four partial products formed, one for each part of the multiplier. Work out the four multiplications that Sara has tried to solve.

a 600 + 60 + 240 + 24

⬜

b 1200 + 120 + 160 + 16

⬜

c 2400 + 240 + 420 + 42

⬜

d 7200 + 480 + 630 + 42

⬜

Date: _____

Lesson 2: **Grid method**

- Use the grid method to multiply numbers to 1000 by 2-digit numbers

1 Partition the numbers in the grids. You don't need to work out the answers.

a   18 × 38 =

Estimate:

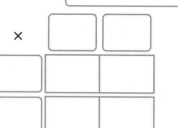

b   46 × 27 =

Estimate:

c   74 × 54 =

Estimate:

d   65 × 93 =

Estimate:

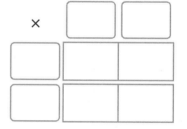

2 For each calculation, estimate first, then use the grid to work out the answer. Show your working. Check your answer with your estimate.

a   22 × 36 =

Estimate:

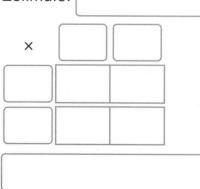

b   45 × 58 =

Estimate:

Number

c 65 × 73 = [ ]

Estimate: [ ]

× [ ] [ ]

[ ] [ ] [ ]

[ ] [ ] [ ]

[ ]

d 76 × 84 = [ ]

Estimate: [ ]

× [ ] [ ]

[ ] [ ] [ ]

[ ] [ ] [ ]

[ ]

3 Use these number cards to write two different word problems where a 2-digit number is multiplied by a 2-digit number.

| 75 | 26 | 48 | 97 | 52 | 33 |

a _____

_____

_____

b _____

_____

_____

Use the grid method to solve your word problems.

Date: _____

Lesson 3: Expanded written method

Number

- Use the expanded written method to multiply numbers to 1000 by 2-digit numbers

1 Complete the calculations. Round and estimate first.

a Estimate: [ ]

		4	3
×		1	7
			1
			0
			0
+		0	0

3	×	7
40	×	7
3	×	10
40	×	10

b Estimate: [ ]

		6	4
×		3	3
			2
			0
			0
+		0	0

4	×	3
60	×	3
4	×	30
60	×	30

2 Complete the calculations. Round and estimate first.

a Estimate: [ ]

		7	6
×		3	7
+			

×	
×	
×	
×	

b Estimate: [ ]

		8	8
×		6	4
+			

×	
×	
×	
×	

Number

3 Complete the calculations. Round and estimate first.

a Estimate: _____

		2	3	7
×			2	4
+				

× _____
× _____
× _____
× _____
× _____
× _____

b Estimate: _____

		3	5	4
×			1	8
+				

× _____
× _____
× _____
× _____
× _____
× _____

c Estimate: _____

	6	1	3
×		4	7
+			

× _____
× _____
× _____
× _____
× _____
× _____

d Estimate: _____

	8	6	4
×		6	3
+			

× _____
× _____
× _____
× _____
× _____
× _____

Date: _____

Lesson 4: **Real-life problems**

Number

- Solve problems involving multiplication of numbers to 1000 by 2-digit numbers

You will need
- paper
- coloured pencil

1 All the seats on two coaches have been booked.

Which coach will make more money, SpeedyCoach or FastCoach?

Answer: _____

Working out

SPEEDYCOACH

Tickets cost $88

The bus has 9 rows of seats.

Each row has 4 seats.

FASTCOACH

Tickets cost $76

The bus has 7 rows of seats.

Each row has 6 seats.

2 Solve the problems. Use paper to show your working. Remember to estimate first.

a A factory puts 328 pencils in each box.

How many pencils will there be in 56 boxes? ☐ pencils

b There are 67 bags of counters. Each bag contains 813 counters.

How many counters are there in all? ☐ counters

c A chocolate factory makes 642 bars of chocolate each day. How many

bars of chocolate will the factory make in 34 days? ☐ bars

d Laptops are boxed up and arranged in stacks of 7 waiting to be delivered. There are 4 stacks. If each laptop is priced at $924, how much money will the company receive once all are sold?

$ [ ]

3 Amir plays a dartboard game. He must only hit the pairs of sectors (one inner and one outer) that multiply to give a product between 32 500 and 33 500. Help him by colouring the sectors where he should aim a dart. Use paper to show your working. One sector has been completed for you.

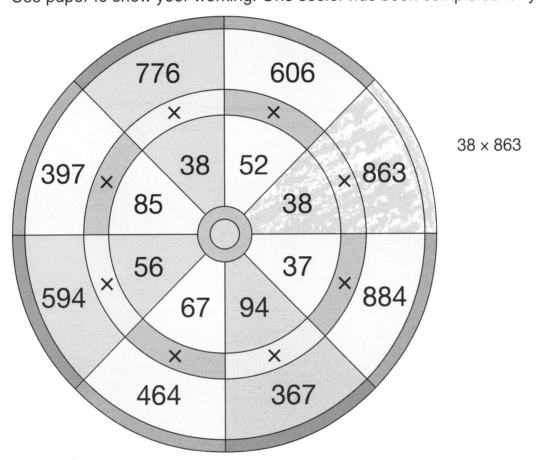

38 × 863

4 A lorry driver needs to calculate the mass of items on his truck. 37 blue crates have been loaded, each with a mass of 414 kg, 48 green crates have been loaded, each with a mass of 377 kg, and 64 red crates have been loaded, each with a mass of 633 kg. What is the combined mass of all the crates? Use paper to show your working.

Estimate: [ ] [ ] kg

Date: _____

Lesson 1: **Working with place value (1)**

- Use place value to divide numbers to 100 by 1-digit numbers

1 Use the empty number lines to work out the answers to these division questions.

Example: 32 ÷ 2 = 16

ten 2s　　　six 2s

0　　　20　　　32

a 　0　　38 ÷ 2 = ☐

b 0　　45 ÷ 3 = ☐

c 0　　84 ÷ 6 = ☐

d 0　　80 ÷ 5 = ☐

2 For each division, write two multiplication facts to help find the answer.

a 60 ÷ 4 =

☐ × 4 = ☐

☐ × 4 = ☐

So 60 ÷ 4 = ☐

b 90 ÷ 5 =

☐ × 5 = ☐

☐ × 5 = ☐

So 90 ÷ 5 = ☐

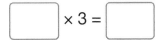
Number

c 78 ÷ 6 =

☐ × 6 = ☐

☐ × 6 = ☐

So 78 ÷ 6 = ☐

d 42 ÷ 3 =

☐ × 3 = ☐

☐ × 3 = ☐

So 42 ÷ 3 = ☐

 Explain how you would find the answer to these division questions. Express any remainder as a fraction of the divisor.

a How many 3s are in 51?

_____

b How many 4s are in 78?

_____

c How many 5s are in 79?

_____

 Think about these calculations: 72 ÷ 3   96 ÷ 6   92 ÷ 4   95 ÷ 5

Which of these calculations gives the smallest quotient? The largest quotient? How do you know?

_____

_____

5 Toby, Olivia and Krishna are dividing 2-digit numbers by 1-digit numbers. Toby's answer is 15. Olivia's answer is 18. Krishna's answer is 14. What might their division questions be?

Toby	Olivia	Krishna

Date: _____

Lesson 2: **Working with place value (2)**

- Using place value to divide numbers
  to 100 by 1-digit numbers

You will need
- place value counters
- paper

1 Write in the missing numbers and find
the quotient.

a $48 \div 3 =$ [÷] + [÷] = [+] = [ ]

b $78 \div 6 =$ [÷] + [÷] = [+] = [ ]

c $91 \div 7 =$ [÷] + [÷] = [+] = [ ]

d $95 \div 5 =$ [÷] + [÷] = [+] = [ ]

2 Use place value counters to model each division calculation. Find the
quotient by completing the steps. Estimate the answer first.

a $92 \div 4 =$ [ ]   Estimate: [ ]

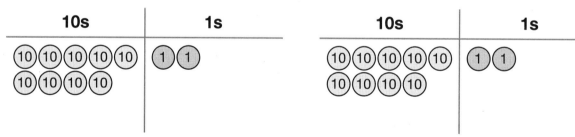

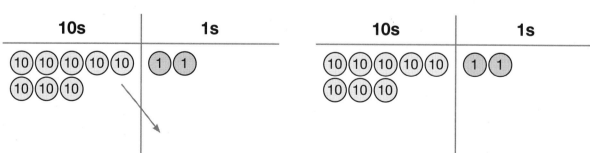

Number

b $82 \div 3 =$ ☐   Estimate: ☐

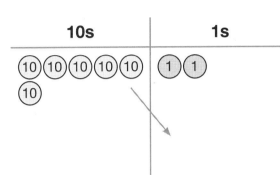

10s	1s
⑩⑩⑩⑩⑩ ⑩⑩⑩	① ①

10s	1s
⑩⑩⑩⑩⑩ ⑩⑩⑩	① ①

10s	1s
⑩⑩⑩⑩⑩ ⑩	① ①

10s	1s
⑩⑩⑩⑩⑩ ⑩	① ①

3   For each calculation, write the number of tens that need to be exchanged for ones.

a $56 \div 4$ ☐ ten(s)   **b** $81 \div 3$ ☐ ten(s)

c $76 \div 4$ ☐ ten(s)   **d** $54 \div 3$ ☐ ten(s)

4   Solve the calculations and problems using any method you prefer, mental or written. Use paper to show your working. Remember to estimate first.

a $64 \div 4 =$ ☐   **b** $51 \div 3 =$ ☐   **c** $84 \div 6 =$ ☐

d $83 \div 6 =$ ☐   **e** $98 \div 8 =$ ☐   **f** $94 \div 7 =$ ☐

g $77 \div 4 =$ ☐   **h** $76 \div 3 =$ ☐   **i** $89 \div 6 =$ ☐

j   91 penguins need to be put equally into 7 pools.

How many will go in each pool? ☐ penguins

k   A supermarket has 8 shelves on which to evenly stack 96 tins of soup.

How many tins can go on each shelf? ☐ tins

Date: _____   ☺ 😐 ☹

Lesson 3: **Expanded written method**

- Use the expanded written method to divide numbers to 100 by 1-digit numbers

1 Write the multiples of 8 from 8 to 80.

Use your list to write the multiple of 8 that comes **before** each number.

a ☐ 27   **b** ☐ 22   **c** ☐ 77

d ☐ 51   **e** ☐ 69   **f** ☐ 35

 2 Estimate first, then use the expanded written method of division to work out the answer to each calculation. Remember to use multiplication to check the results of the division.

There are no remainders in these answers.

a 87 ÷ 3 = ☐

Estimate: ☐

 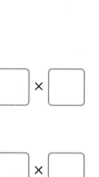

b 98 ÷ 7 = ☐

Estimate: ☐

 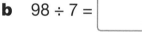 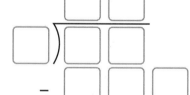 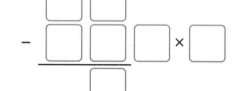

Number

There are remainders in these answers.

c $81 \div 6 =$ ☐

Estimate: ☐

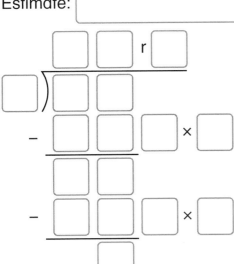

d $91 \div 4 =$ ☐

Estimate: ☐

---

3 Solve the word problems. Use the expanded written method of division. Estimate each answer first.

a A shop has 534 shirts. If the shirts are stored in piles of 6, how many piles are there?

b A school has $547 to buy new books. Each book costs $8. How many books can the school buy? How much money is left over?

Date: _____

Lesson 4: **Real-life problems**

> • Solve problems involving division of numbers to 100 by 1-digit numbers

1 Complete the division problems using any method you prefer.

a $44 \div 4 = \boxed{}$ **b** $39 \div 3 = \boxed{}$ **c** $65 \div 5 = \boxed{}$

d $96 \div 8 = \boxed{}$ **e** $84 \div 7 = \boxed{}$ **f** $99 \div 9 = \boxed{}$

g $84 \div 6 = \boxed{}$ **h** $48 \div 3 = \boxed{}$ **i** $85 \div 5 = \boxed{}$

j $76 \div 4 = \boxed{}$ **k** $96 \div 6 = \boxed{}$ **l** $93 \div 3 = \boxed{}$

m $91 \div 7 = \boxed{}$ **n** $95 \div 5 = \boxed{}$ **o** $84 \div 3 = \boxed{}$

2 These word problems do not have numbers in them, but you should be able to tell what to do if there is a remainder.

Write 'Round up' or 'Round down' for each problem.

Example: There are $\boxed{}$ biscuits to be put into packets.

Each packet holds $\boxed{}$ biscuits.

How many full packets can you make? $\boxed{\text{Round down}}$

a A baker is making cakes. Each cake tray has space for $\boxed{}$ cakes.

The baker has made enough mixture to bake $\boxed{}$ cakes altogether.

How many full cake trays will she have? $\boxed{}$

b A class of $\boxed{}$ learners is told to work in groups of $\boxed{}$.

Some learners may need to work in a smaller group.

How many groups will there be? $\boxed{}$

3 Solve each of these word problems. Decide whether to round the answer up or down to answer the problem fully.

a Tinaya has to take 5 ml of cough medicine every day. The bottle contains 67 ml of medicine. How many days' worth of medicine does Tinaya have?

b A farmer needs to build a fence 55 metres long. Fencing is sold in 3-metre reels. How many reels must the farmer buy?

4 Malik divides a mystery number by 2 and says that he has a remainder of 1. Jessica says that Malik's mystery number must have been odd.

Is Jessica right? ☐ Explain your answer.

5 Malik divides a mystery number by 5 and says that he has a remainder of 3. Jessica says that there are only two possible ones digits that Malik's mystery number must end with.

What are the digits? ☐ or ☐ Explain how Jessica knows this.

Date: _____

Number

Lesson 1: **Working with place value (1)**

• Use place value to divide numbers to 1000 by 1-digit numbers

1 Use known division facts to solve these problems.

a 6 ÷ 2 = ☐

b 60 ÷ 2 = ☐

c 600 ÷ 2 = ☐

d 12 ÷ 3 = ☐

e 120 ÷ 3 = ☐

f 1200 ÷ 3 = ☐

g 28 ÷ 7 = ☐

h 280 ÷ 7 = ☐

i 2800 ÷ 7 = ☐

2 Use known number facts or partitioning to complete each division.

a 250 ÷ 5 = ☐

b 306 ÷ 3 = ☐

c 260 ÷ 2 = ☐

d 780 ÷ 6 = ☐

e 108 ÷ 4 = ☐

f 246 ÷ 3 = ☐

g 357 ÷ 7 = ☐

h 416 ÷ 8 = ☐

i 558 ÷ 9 = ☐

3 Use the mental partitioning method of division to work out each calculation. Show your working. Then using the Code on page 71, write the letter for each answer in the box shown by the arrow to reveal the secret phrase.

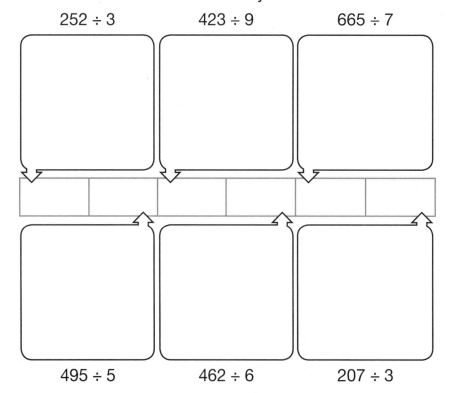

252 ÷ 3   423 ÷ 9   665 ÷ 7

495 ÷ 5   462 ÷ 6   207 ÷ 3

Number

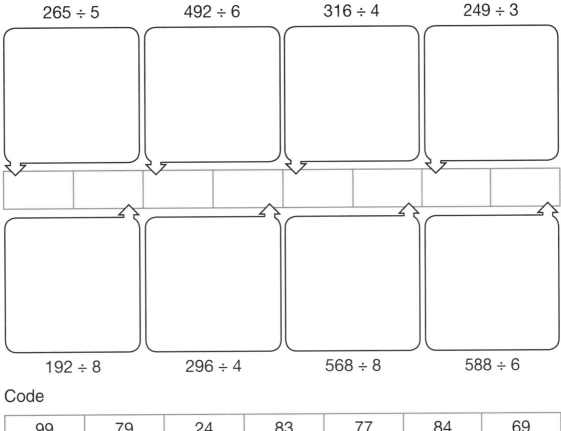

265 ÷ 5   492 ÷ 6   316 ÷ 4   249 ÷ 3

192 ÷ 8   296 ÷ 4   568 ÷ 8   588 ÷ 6

Code

99	79	24	83	77	84	69
E	T	T	G	T	M	L

71	47	53	74	82	95	98
E	N	S	A	R	A	Y

4 Some of the inputs and outputs for each function machine are missing.

Work out the missing numbers.

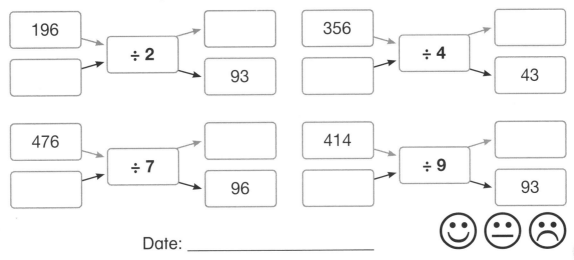

Date: _____

Lesson 2: **Working with place value (2)**

Number

- Use place value to divide numbers to 1000 by 1-digit numbers

You will need
- place value counters
- paper

1 Fill in the missing numbers and find the quotients.

a 159 ÷ 3 = [ ] ÷ [ ] + [ ] ÷ [ ] = [ ] + [ ] = [ ]

b 378 ÷ 6 = [ ] ÷ [ ] + [ ] ÷ [ ] = [ ] + [ ] = [ ]

c 294 ÷ 7 = [ ] ÷ [ ] + [ ] ÷ [ ] = [ ] + [ ] = [ ]

d 736 ÷ 8 = [ ] ÷ [ ] + [ ] ÷ [ ] = [ ] + [ ] = [ ]

2 Use place value counters to model each division calculation. Find the quotient by completing the steps. Estimate the answer first.

a 144 ÷ 4 = [ ]   Estimate: [ ]

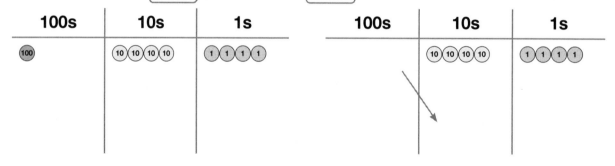

Number

b $255 \div 7 =$ [ ]   Estimate: [ ]

100s	10s	1s
(100)(100)	(10)(10)(10)(10)(10)	(1)(1)(1)(1)(1)

100s	10s	1s
	(10)(10)(10)(10)(10)	(1)(1)(1)(1)(1)

100s	10s	1s
	(10)(10)(10)(10)(10)	(1)(1)(1)(1)(1)

100s	10s	1s
	(10)(10)(10)(10)(10)	(1)(1)(1)(1)(1)

 3 Solve the calculations. Use any method you prefer, mental or written.
Use paper to show your working. Remember to estimate first.
These divisions have no remainders.

a $292 \div 4 =$ [ ]   **b** $498 \div 6 =$ [ ]   **c** $297 \div 3 =$ [ ]

d $465 \div 5 =$ [ ]   **e** $161 \div 7 =$ [ ]   **f** $536 \div 8 =$ [ ]

These divisions have remainders.

g $769 \div 8 =$ [ ]   **h** $458 \div 7 =$ [ ]   **i** $535 \div 6 =$ [ ]

j $832 \div 9 =$ [ ]   **k** $346 \div 4 =$ [ ]   **l** $409 \div 8 =$ [ ]

 4 Find five 3-digit numbers that give a remainder of 3 when divided by 7.
Each 3-digit number must have a different hundreds digit. Write the
numbers and describe the strategy you used. [ ] [ ] [ ]

_____

_____

Date: _____

Lesson 3: **Expanded written method**

- Use the expanded written method to divide numbers to 1000 by 1-digit numbers

1 Write the multiples of 60, from 60 to 600.

Use your list to write the multiple of 60 that comes **before** each number.

a ☐ 132   **b** ☐ 318   **c** ☐ 264

d ☐ 216   **e** ☐ 462   **f** ☐ 510

2 Estimate first, then use the expanded written method of division to work out the answer to each calculation. Remember to use multiplication to check the results of the division.

These divisions have no remainders.

a 258 ÷ 6 = ☐

Estimate: ☐

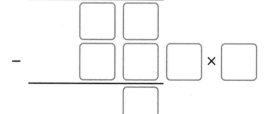

b 539 ÷ 7 = ☐

Estimate: ☐

These divisions have remainders.

c 187 ÷ 8 = ⬚

Estimate: ⬚

d 249 ÷ 9 = ⬚

Estimate: ⬚

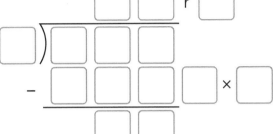
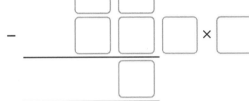

3 Solve the word problems. Estimate each answer first.

a 5341 fans attend a football match. They are seated in equal numbers in 7 areas of the stadium. How many fans are in each area?

b 5168 oranges are packed in 8 large crates. Each crate has an equal number of oranges. How many oranges are in each crate?

c 6984 kg of soil is divided equally between 9 gardens. How much soil will there be in each garden?

Date: _____

Lesson 4: **Real-life problems**

Number

- Solve problems involving division of numbers to 1000 by 1-digit numbers

You will need
- paper

1 Complete the division problems using any method you prefer.

a 147 ÷ 3 =

b 236 ÷ 4 =

c 455 ÷ 5 =

d 204 ÷ 6 =

e 294 ÷ 7 =

f 720 ÷ 8 =

g 351 ÷ 9 =

h 186 ÷ 3 =

i 316 ÷ 4 =

j 415 ÷ 5 =

k 498 ÷ 6 =

l 616 ÷ 7 =

m 704 ÷ 8 =

n 369 ÷ 9 =

o 584 ÷ 8 =

2 Solve each of these word problems. Decide whether to round any remainders up or down. Use paper to show your working.

a Nathan's book has 184 pages. He reads 7 pages every day.

How many days does it take him to finish the book?

b Ella has 477 magazines to deliver to shops. They are collected together in packs of 8.

How many complete packs does she have?

c A gardener has $672. He wants to buy plants that cost $9 each.

How many can he buy?

3 Solve each of these word problems. Use the expanded written method of division. Decide whether to round any answers with remainders up or down.

a A shop has 525 eggs to sell. The eggs are arranged in boxes of 6. How many complete boxes are there?

Number

b Sanaa wants to buy 515 cupcakes for a party. The cupcakes come in boxes of 7. How many boxes should she buy?

c An office manager has $621 to buy new lightbulbs. Each lightbulb costs $8. How many bulbs can he buy? How much money is left over?

d 9 people can be seated at each table in a restaurant. There are 790 guests. How many tables are needed? How many guests will be on a table with fewer than 9 people?

4 Answer the word problems. Decide whether to round any answers with remainders up or down. Use paper to show your working.

a 2420 people want to ride a roller coaster. The cars hold 7 people each.

How many cars will be full?

b 1760 large boxes are loaded into vans that each have a capacity of 9 boxes.

How many vans will be needed to transport all of the boxes?

c 3250 kg of sand needs to be shipped in containers that can each carry a maximum mass of 950 kg.

How many containers will the sand fill?

Date: _____

Lesson 1: **Decimal place value**

- Explain the value of the tenths and hundredths digits in decimals

1 In each box, write the number the arrow points to.

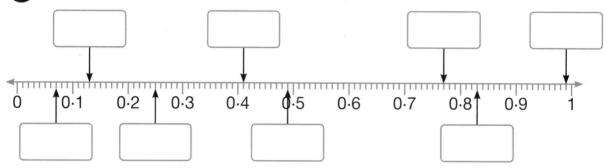

2 Write the value of each underlined digit.

Example: 3·84 | 8 tenths or 0.8 |

a   0·2<u>7</u>

b   3·<u>4</u>8

c   <u>7</u>·19

d   0·<u>0</u>2

e   9·0<u>8</u>

f   <u>1</u>6·33

g   −4·<u>3</u>4

h   −12·0<u>5</u>

3 Write the decimal that is equivalent to each fraction.

a $\frac{3}{10}$ =

b $\frac{7}{10}$ =

c $\frac{5}{10}$ =

d $\frac{8}{10}$ =

e $\frac{33}{100}$ =

f $\frac{47}{100}$ =

g $\frac{79}{100}$ =

h $\frac{11}{100}$ =

i $\frac{70}{100}$ =

j $\frac{1}{100}$ =

k $\frac{61}{100}$ =

l $\frac{50}{100}$ =

Number

4 Write an example of a number with:

a 4 tenths between 5 and 6 ☐

b 7 hundredths between 3·2 and 3·3 ☐

c 78 hundredths between 14 and 15 ☐

d negative 9 tenths between −9 and −10 ☐

e negative 2 hundredths between −3·5 and −3·6 ☐

f negative 99 hundredths between −99 and −100 ☐

Number

5 Find the total in each box. Write the answer in numerals.

| nine |
| 7 hundredths |
| 3 tenths |

a ☐

| 4 | 0·02 |
| 0·8 | 50 |

b ☐

5	200
0·1	
30	0·0

c ☐

five hundred	
	seventy
	seven
2 tenths	
	6 hundredths

d ☐

eight hundred	
	1 hundredth
six	
	forty
four thousand	

e ☐

| 90 | 0·9 | 0·9 |
| 900 | 9000 | 9 |

f ☐

Date: _____

79

Number

Lesson 2: **Composing and decomposing decimals**

• Compose and decompose decimals

1 Draw lines to match the decimals to their decomposed forms.

| 6·78 |

| 8·76 |

| 7·86 |

| 60·87 |

| 8 + 0·7 + 0·06 |

| 6 + 0·7 + 0·08 |

| 6 tens + 8 tenths + 7 hundredths |

| 7 ones + 8 tenths + 6 hundredths |

 2 Decompose the numbers by place value.

a 0·67 = ☐ + ☐

b 0·33 = ☐ + ☐

c 0·51 = ☐ + ☐

d 4·89 = ☐ + ☐ + ☐

e 9·11 = ☐ + ☐ + ☐

f 6·06 = ☐ + ☐

g 5·55 = ☐ + ☐ + ☐

h 3·52 = ☐ + ☐ + ☐

Number

3 Complete the sentences.

17·92 is composed from 10, 7, 0·9 and 0·02.

a 15·34 is composed from

b 25·29 is composed from

c 30·75 is composed from

d 41·02 is composed from

e 623·22 is composed from

f 809·06 is composed from

4 Find four different ways to decompose each number.

Example: 4·78

i 4 + 0·7 + 0·08

ii 478 hundredths

iii 4 ones and 78 hundredths

iv 47 tenths and 8 hundredths

a 8·36

i

ii

iii

iv

b 9·15

i

ii

iii

iv

Date: _____

Lesson 3: **Regrouping decimals**

- Regroup decimals to help with calculations

1 Decompose each number by place value.

$4·78 = 4 + 0·7 + 0·08$

a $6·63 = $ ☐ + ☐ + ☐

b $9·14 = $ ☐ + ☐ + ☐

c $8·88 = $ ☐ + ☐ + ☐

d $17·29 = $ ☐ + ☐ + ☐ + ☐

e $32·52 = $ ☐ + ☐ + ☐ + ☐

f $50·44 = $ ☐ + ☐ + ☐ + ☐

g $86·07 = $ ☐ + ☐ + ☐ + ☐

2 Write the decimal.

a 6 ones, 2 tenths and 9 hundredths ☐

b 85 ones, 5 tenths and 3 hundredths ☐

c 38 tenths and 4 hundredths ☐

d 999 hundredths ☐

3 Decompose each number in four different ways.

a

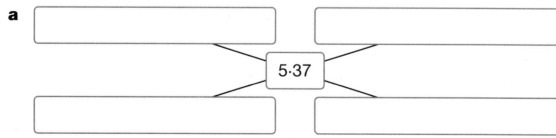

5·37

Number

b

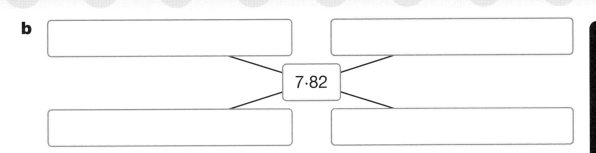

7·82

c

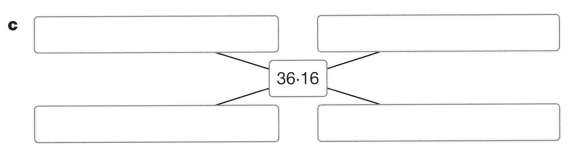

36·16

d

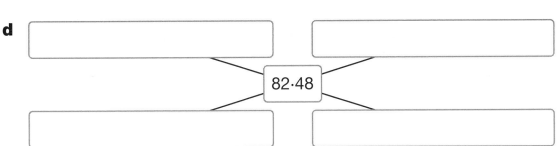

82·48

4 How would you use decomposition and regrouping to solve these subtraction problems?

CLUE

Think:
30 + 62 tenths + 8 hundredths
– 10 + 27 tenths + 3 hundredths

a 7·65 – 2·37

b 8·76 – 3·58

c 36·28 – 12·73

Date: _____

😊 😐 ☹

83

Lesson 4: **Comparing and ordering decimals**

• Compare and order decimals

0 0·2 0·4 0·6 0·8 1 1·2 1·4 1·6 1·8 2 2·2 2·4 2·6 2·8 3 3·2 3·4 3·6 3·8 4 4·2 4·4 4·6 4·8 5

Use the number line to answer the questions.

1 Order each set of decimals, starting with the smallest.

a 1·4, 1, 0·4 [   ]   **b** 3·6, 2, 3·2 [   ]

c 4·8, 4·2, 2·4 [   ]   **d** 2·4, 2·8, 2·2 [   ]

2 Which decimal number is halfway between:

a 1·2 and 1·8? [   ]   **b** 3·5 and 4·1? [   ]

c 0·1 and 0·9? [   ]   **d** 2·4 and 3·4? [   ]

3 Write the correct symbol, < or >, to compare each pair of decimals.

a 0·4 [   ] 0·8   **b** 0·3 [   ] 0·1   **c** 1·7 [   ] 1·9

d 0·5 [   ] 0·2   **e** 0·6 [   ] 0·7   **f** 2·5 [   ] 2·4

g 5 [   ] 5·1   **h** 11·1 [   ] 11·3   **i** 91·4 [   ] 91·9

j 16·4 [   ] 16·5   **k** 14·5 [   ] 15·4   **l** 10·1 [   ] 1·01

4 Write the correct symbol, < or >, to compare each pair of measurements.

a 0·5 m [   ] 0·3 m   **b** 0·9 kg [   ] 0·8 kg

c 0·7 cm [   ] 0·4 cm   **d** 0·2 km [   ] 0·7 km

e 0·4 *l* [   ] 0·1 *l*   **f** 4·2 mm [   ] 4·3 mm

g 12·2 g [   ] 12 g   **h** 50·7 m*l* [   ] 50·1 m*l*

i 15·6° [   ] 16·5°   **j** 25·6 m [   ] 24·8 m

5 Toby was asked to arrange sets of numbers in order, from greatest to smallest.

Has Toby made any mistakes? If so, how would you correct the order?

a   0·7, 0·5, 0·4, 0·3, 0·1   **b**   7·3, 6·9, 6·5, 6·6, 6·1

c   8·8, 8·6, 8·4, 8, 8·1   **d**   26·6, 26·2, 25·8, 24·7, 23·2

e   65·5, 65·3, 66·1, 65·1, 64·7   **f**   97·6, 96·7, 95·6, 94·9, 91·2

6 Complete the decimals to make each statement true.

a $2·4 < 2·\boxed{}$ **b** $210·5 > 210·\boxed{}$ **c** $13·\boxed{} > 13·7$

d $199·8 < 199·\boxed{}$ **e** $20·5 < 20·\boxed{}$ **f** $105·\boxed{} > 105$

g $10·2 > 1\boxed{}·\boxed{}$ **h** $400·7 < 400·\boxed{}$

i $37·\boxed{} < 37·5$ **j** $354·6 < 35\boxed{}·\boxed{}$

7 Clara says,

I'm thinking of three decimals that each have one decimal place. All of the numbers are between 87 and 88. The first number has three more tenths than the second number. The second number has five fewer tenths than the third number.

Generalise and write all the different possibilities for the three decimals, ordering each set from smallest to greatest.

Date: _____

Number

Lesson 1: **Multiplying and dividing by 10, 100 and 1000**

- Multiply and divide whole numbers by 10, 100 and 1000

You will need

- yellow and blue coloured pencils

1 Fill in the missing outputs.

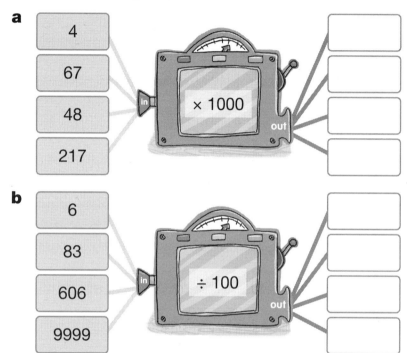

a

| 4 |
| 67 |
| 48 |
| 217 |

× 1000

b

| 6 |
| 83 |
| 606 |
| 9999 |

÷ 100

 2 Complete the calculations.

a 9 × 10 = ☐

b 9 × 100 = ☐

c 9 × 1000 = ☐

d 820 ÷ 10 = ☐

e 820 ÷ 100 = ☐

f 820 ÷ 1000 = ☐

g 76 × 100 = ☐

h 76 × 1000 = ☐

i 76 × 10 = ☐

j 4530 ÷ 100 = ☐

k 4530 ÷ 1000 = ☐

l 4530 ÷ 10 = ☐

Number

3 The number to be multiplied is in bold. The product of the multiplication is below it. Colour yellow the squares containing numbers that have been multiplied by 10. Colour blue the squares containing numbers that have been multiplied by 100.

55	484	9	123	237
5500	48 400	90	12 300	2370
6	**5054**	**708**	**99**	**106**
600	505 400	7080	9900	10 600
22	**404**	**10 000**	**303**	**1**
220	40 400	1 000 000	3030	100
666	**1111**	**37**	**4043**	**8089**
66 600	111 100	3700	40 430	808 900

4 Complete the table by converting the amount in cents to dollars.

cents	4	70	58	204	3009
$					

5 Work out the cost of each set of items.

a 10 costing $2 each.

b 100 costing $3 each.

c 1000 costing $12 each.

d 10 costing $102·50 in total. Price for 1:

e 100 costing $2725 in total. Price for 1:

f 1000 costing $78 890 in total. Price for 1:

Date: _____

Number

Lesson 2: **Multiplying decimals by 10 and 100**

- Multiply decimals by 10 and 100

1 Write the missing numbers in the place value charts.

a

1000s	100s	10s	1s •	$\frac{1}{10}$s	$\frac{1}{100}$s	
			2 •	3		
			•			× 10
			•			× 100

b

1000s	100s	10s	1s •	$\frac{1}{10}$s	$\frac{1}{100}$s	
		4	8 •	4		
			•			× 10
			•			× 100

c

1000s	100s	10s	1s •	$\frac{1}{10}$s	$\frac{1}{100}$s	
		2	3 •	7	9	
			•			× 10
			•			× 100

2 a How many millimetres is 5·7 cm? ☐ mm

b How many centimetres is 0·63 metres? ☐ cm

Remember!

1 cm = 10 mm

1 m = 100 cm

3 Fill in the missing outputs.

a

26·22

50·6

88·99

125·2

in   × 10   out

88

b

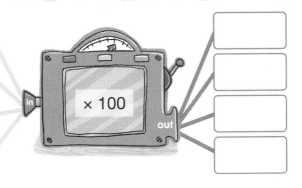

| 64·09 |
| 10·8 |
| 404·04 |
| 1·1 |

× 100

4 Draw a line to match each mass to its description.

| 100 times heavier than 1·07 kg | 100 times heavier than 0·17 kg | 100 times heavier than 0·71 kg | 10 times heavier than 1·07 kg | 100 times heavier than 1·7 kg |

| 71 kg | 10·7 kg | 170 kg | 107 kg | 17 kg |

5 Work out the answers to these addition problems. Show your working.

a | 7 × 100 | + | 70 × 100 | + | 0·7 × 10 | + | 0·07 × 10 | = | |

b | 0·5 × 100 | + | 5000 × 10 | + | 50 × 100 | + | 0·05 × 10 | = | |

c | 2·03 × 100 | + | 20·3 × 10 | + | 203 × 100 | + | 0·23 × 10 | = | |

Date: _____

Number

Lesson 3: **Dividing by 10 and 100**

• Divide decimals by 10 and whole numbers by 10 and 100

1 Write the missing numbers in the place value charts.

a

100s	10s	1s •	$\frac{1}{10}$s	$\frac{1}{100}$s
		2 •	9	

÷ 10

100s	10s	1s •	$\frac{1}{10}$s	$\frac{1}{100}$s
		•		

b

100s	10s	1s •	$\frac{1}{10}$s	$\frac{1}{100}$s
	8	0 •	2	

÷ 10

100s	10s	1s •	$\frac{1}{10}$s	$\frac{1}{100}$s
		•		

c

100s	10s	1s •	$\frac{1}{10}$s	$\frac{1}{100}$s
	1	8 •		

÷ 100

100s	10s	1s •	$\frac{1}{10}$s	$\frac{1}{100}$s
		•		

d

100s	10s	1s •	$\frac{1}{10}$s	$\frac{1}{100}$s
3	6	1 •		

÷ 100

100s	10s	1s •	$\frac{1}{10}$s	$\frac{1}{100}$s
		•		

2 **a** How many cm is 4·5 mm? ☐ cm

b How many cm is 50·1 mm? ☐ cm

Remember!

10 mm = 1 cm

3 Fill in the missing outputs.

a

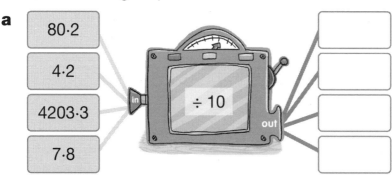

80·2

4·2

4203·3

7·8

÷ 10

in

out

Number

b

10

103

3

1450

÷ 100

in

out

4 Draw a line to match each mass to its description.

| 100 times lighter than 62 kg | 100 times lighter than 6 kg | 10 times lighter than 6·1 kg | 10 times lighter than 62 kg | 100 times lighter than 621 kg |

| 0·61 kg | 0·62 kg | 6·21 kg | 0·06 kg | 6·2 kg |

5 Work out the answers to these addition problems. Show your working.

a | 800 ÷ 100 | + | 8 ÷ 10 | + | 0·8 ÷ 10 | + | 800 ÷ 10 | = | |

b | 700 ÷ 10 | + | 7 ÷ 10 | + | 70 ÷ 10 | + | 7 ÷ 100 | = | |

c | 7500 ÷ 100 | + | 75 ÷ 100 | + | 2 ÷ 100 | + | 2 ÷ 10 | = | |

Date: _____

Number

Lesson 4: **Rounding decimals to the nearest whole number**

• Round decimals to the nearest whole number

1 Round each number to the nearest whole number.

a
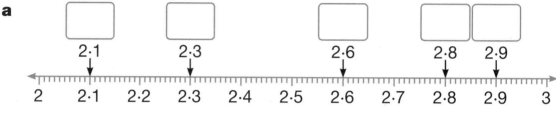

b
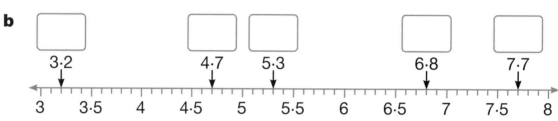

2 Write the two whole numbers on either side of each decimal.

Then draw a ring around the number that the decimal rounds to.

a [ ] 4·3 [ ]   **b** [ ] 8·6 [ ]

c [ ] 1·4 [ ]   **d** [ ] 6·5 [ ]

e [ ] 14·8 [ ]   **f** [ ] 29·4 [ ]

g [ ] 53·2 [ ]   **h** [ ] 88·9 [ ]

3 Round each measurement to the nearest whole unit.

a 0·4 kg [ ] kg   **b** 2·5 m [ ] m   **c** 7·8 *l* [ ] *l*

d 4·3 km [ ] km   **e** 11·7 g [ ] g   **f** 6·9 cm [ ] cm

g 0·4 seconds [ ] seconds   **h** 23·6 m [ ] m

i 55·5 kg [ ] kg   **j** 3·1 *l* [ ] *l*

Number

 Draw a ring around the numbers that round up to the nearest whole number.

5·8   2·5   6·2   18·7   1·3   11·6   7·9   8·1

 Draw a ring around the numbers that round down to the nearest whole number.

3·6   18·3   4·7   21·4   17·8   5·5   0·9   1·2

 Round the prices to the nearest dollar.

a $3·10 [ ]   b $5·50 [ ]

c $8·70 [ ]   d $41·20 [ ]

e $29·90 [ ]   f $55·50 [ ]

7 Write all the numbers with one decimal place that round to:

2 a   5 [ ]

b   8 [ ]

c   27 [ ]

d   99 [ ]

8 Write the two tenths either side of each decimal.

Then draw a ring around the number that the decimal rounds to.

The first one has been done for you.

a   [0·2]   0·27   (⌒0·3)   b   [ ]   0·44   [ ]

c   [ ]   0·75   [ ]   d   [ ]   2·38   [ ]

e   [ ]   6·12   [ ]   f   [ ]   4·82   [ ]

g   [ ]   5·29   [ ]   h   [ ]   7·01   [ ]

Date: _____

Lesson 1: **Fractions as division**

Number

> • Understand that a fraction can be represented by a division of the numerator by the denominator

1 Draw a line to match each fraction to its division statement.

$\frac{1}{4}$ •   • 1 divided by 5

$\frac{1}{3}$ •   • 1 divided by 10

$\frac{1}{10}$ •   • 1 divided by 4

$\frac{1}{5}$ •   • 1 divided by 2

$\frac{1}{2}$ •   • 1 divided by 8

$\frac{1}{8}$ •   • 1 divided by 3

2 Fill in the table to represent each fraction as a division.

Fraction	Fraction as a division
$\frac{3}{4}$	$3 \div 4$
$\frac{2}{5}$	
$\frac{4}{5}$	
$\frac{3}{10}$	
$\frac{7}{10}$	
$\frac{9}{100}$	
$\frac{33}{100}$	
$\frac{97}{100}$	

3 Atique divides a set of beads into 5 equal shares.

What fraction of beads is in each share?

How would you represent this fraction as a division?

Number

4 Chen divides a packet of pens into 10 equal shares.

What fraction of pens is in each share?

How would you represent this fraction

as a division?

5 Taman takes a string that is 9 m in length and divides it into 10 equal pieces.

What fraction describes each piece of string?

How would you represent this fraction

as a division?

6 Georgia says:

To find a quarter of a number, as well as dividing by 4, I can also divide by 2 twice.

a Use **two different** numbers from the box to test whether Georgia's idea works.

48	84	96	124	56

b Does Georgia's idea work? Yes / No

7 You can find one-tenth of a number without dividing by 10. It involves dividing by two different numbers. Show how. Explain how to find one-tenth without dividing by 10.

Date: _____

Lesson 2: **Improper fractions and mixed numbers**

- Recognise improper fractions and mixed numbers

You will need
- coloured pencil

1 Show each mixed number.

a Colour $1\frac{1}{2}$.

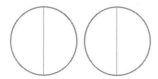

b Colour $1\frac{2}{3}$.

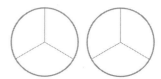

c Colour $1\frac{4}{5}$.

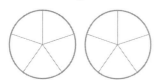

d Colour $1\frac{7}{10}$.

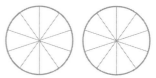

2 Classify each fraction in the correct set.

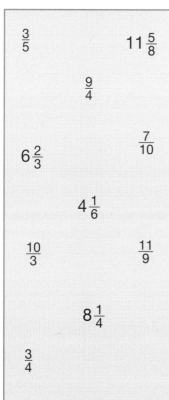

$\frac{3}{5}$ $11\frac{5}{8}$

$\frac{9}{4}$

$\frac{7}{10}$

$6\frac{2}{3}$

$4\frac{1}{6}$

$\frac{10}{3}$ $\frac{11}{9}$

$8\frac{1}{4}$

$\frac{3}{4}$

Proper fractions

Improper fractions

Mixed numbers

Number

3 Write each of these as mixed numbers.

a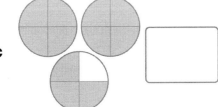

b

c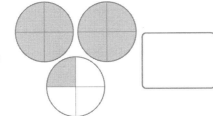

d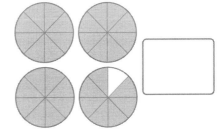

e

f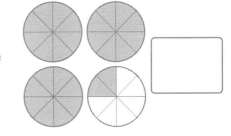

4 Using each of the digits 1, 2 and 3 once only, how many different mixed numbers and improper fractions can you make?

Check the definition of an improper fraction.

Date: _____

Number

Lesson 3: **Converting improper fractions and mixed numbers**

• Convert between improper fractions and mixed numbers

You will need
• coloured pencil

1 Express each diagram as an improper fraction and a mixed number.

a

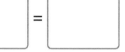

$\square = \square$

b
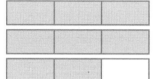

$\square = \square$

c
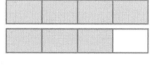

$\square = \square$

d

$\square = \square$

e
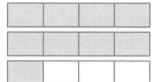

$\square = \square$

f
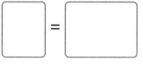

$\square = \square$

g
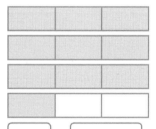

$\square = \square$

h
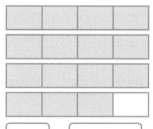

$\square = \square$

i
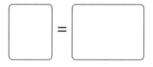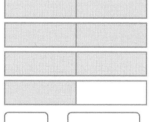

$\square = \square$

2 Shade the circle models to help you convert each improper fraction to a mixed number. Shade as many circles as you need.

a $\frac{13}{4} = $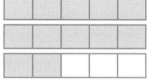

b $\frac{10}{3} = $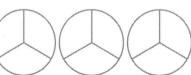

Number

c $\frac{27}{5} =$ [ ]

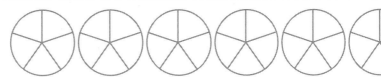

d $\frac{47}{8} =$ [ ]

3 Convert the mixed numbers to improper fractions.

a $3\frac{2}{3}$

b $2\frac{1}{2}$

c $4\frac{3}{4}$

d $8\frac{2}{5}$

e $9\frac{7}{10}$

f $5\frac{11}{12}$

4 Write in the missing numbers.

a $4\dfrac{3}{\boxed{}} = \dfrac{19}{\boxed{}}$

b $6\dfrac{2}{\boxed{}} = \dfrac{50}{\boxed{}}$

c $5\dfrac{3}{\boxed{}} = \dfrac{18}{\boxed{}}$

d $9\dfrac{9}{\boxed{}} = \dfrac{90}{\boxed{}}$

e $8\dfrac{5}{\boxed{}} = \dfrac{85}{\boxed{}}$

f $7\dfrac{1}{\boxed{}} = \dfrac{64}{\boxed{}}$

Date: _____

Number

Lesson 4: **Comparing and ordering fractions**

- Compare and order fractions with the same denominator

You will need
- coloured pencil

1 Write the letter codes of the fractions in the correct order.

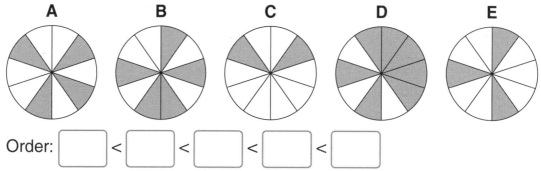

A B C D E

Order: ☐ < ☐ < ☐ < ☐ < ☐

2 Shade each circle to represent the fraction below it. Then use < or > to compare the fractions.

a

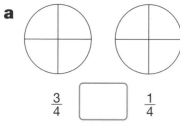

$\frac{3}{4}$ ☐ $\frac{1}{4}$

b

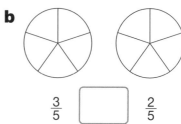

$\frac{3}{5}$ ☐ $\frac{2}{5}$

c

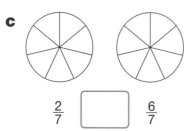

$\frac{2}{7}$ ☐ $\frac{6}{7}$

d

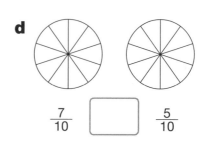

$\frac{7}{10}$ ☐ $\frac{5}{10}$

3 Compare the fractions, using the symbols <, > or =.

a $\frac{7}{8}$ ☐ $\frac{3}{8}$ **b** $\frac{3}{4}$ ☐ $\frac{4}{4}$ **c** $\frac{2}{7}$ ☐ $\frac{4}{7}$

d $\frac{2}{3}$ ☐ $\frac{1}{3}$ **e** $\frac{3}{10}$ ☐ $\frac{1}{10}$ **f** $\frac{5}{10}$ ☐ $\frac{3}{10}$

g 1 ☐ $\frac{9}{9}$ **h** $\frac{4}{6}$ ☐ $\frac{5}{6}$ **i** $\frac{3}{5}$ ☐ $\frac{4}{5}$

4 Write in the missing numbers.

a $\dfrac{\boxed{}}{6} > \dfrac{\boxed{}}{6} > \dfrac{3}{\boxed{}} > \dfrac{\boxed{}}{6}$

b $\dfrac{\boxed{}}{8} < \dfrac{3}{\boxed{}} < \dfrac{\boxed{}}{8} < \dfrac{8}{\boxed{}}$

c $\dfrac{\boxed{}}{10} > \dfrac{\boxed{}}{10} > \dfrac{2}{10} > \dfrac{\boxed{}}{10}$

d $\dfrac{\boxed{}}{9} < \dfrac{4}{\boxed{}} < \dfrac{\boxed{}}{9} < \dfrac{6}{\boxed{}}$

5 Arrange all 16 fractions in the grid so that every row and column is in ascending order.

$\dfrac{7}{24}$ $\dfrac{5}{6}$ $\dfrac{7}{12}$ $\dfrac{7}{8}$ $\dfrac{1}{12}$ $\dfrac{1}{2}$ $\dfrac{13}{24}$ $\dfrac{5}{8}$

$\dfrac{19}{24}$ $\dfrac{1}{6}$ $\dfrac{1}{8}$ $\dfrac{3}{4}$ $\dfrac{11}{12}$ $\dfrac{1}{4}$ $\dfrac{5}{12}$ $\dfrac{3}{8}$

smallest			→ largest
largest			

Date: _____

Number

Lesson 1: **Fractions as operators**

* Understand that proper fractions can act as operators

1 Show the fraction one-tenth on each diagram.

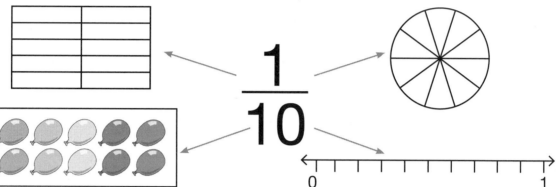

$$\frac{1}{10}$$

0   1

2 Work out these unit fractions.

a $\frac{1}{10}$ of \$70 = ⬚   **b** $\frac{1}{10}$ of 40 km = ⬚   **c** $\frac{1}{3}$ of 99 g = ⬚

d $\frac{1}{4}$ of 240 = ⬚   **e** $\frac{1}{5}$ of 150 kg = ⬚   **f** $\frac{1}{9}$ of 108 km = ⬚

3 Work out these fractions. Show your working out. Remember to estimate first, e.g. $\frac{9}{10}$ is almost 1 so $\frac{9}{10}$ of 460 should be close to 460.

a $\frac{3}{10}$ of \$40 = ⬚

b $\frac{5}{10}$ of 370 km = ⬚

c $\frac{7}{10}$ of 220 g = ⬚

d $\frac{9}{10}$ of 460 = ⬚

e $\frac{3}{100}$ of \$700 = ⬚

f $\frac{9}{100}$ of 900 km = ⬚

Number

4 $\frac{3}{4}$ of the amount of water in each container is poured out. How much water is this?

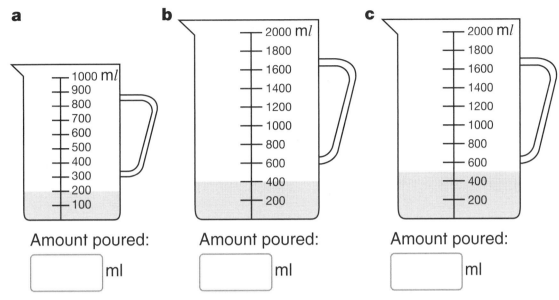

a

b

c

Amount poured:

[ ] ml

Amount poured:

[ ] ml

Amount poured:

[ ] ml

5 Solve these word problems. Show your working out. Estimate the answer first.

a Of the 27 600 supporters at a football game, $\frac{27}{100}$ are wearing hats.

How many supporters are wearing hats? [ ]   Estimate: [ ]

b Of the 13 300 bricks in a wall, $\frac{19}{100}$ are yellow.

How many bricks are yellow? [ ]   Estimate: [ ]

Date: _____

Lesson 2: **Adding and subtracting fractions (1)**

- Add and subtract fractions with the same denominator

You will need
- coloured pencil

1 Shade each circle to show the fraction.

a $\frac{1}{8}$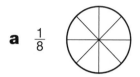

b $\frac{2}{10}$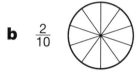

c $\frac{7}{8}$

d $\frac{3}{10}$

e $\frac{2}{9}$

f $\frac{5}{7}$

2 Use the diagrams to add the fractions. Shade the last part of the diagram. Then complete the number sentence.

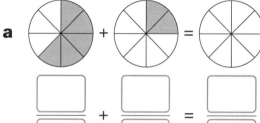

a

b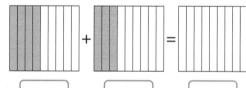

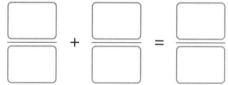

c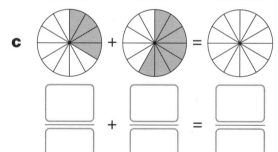

d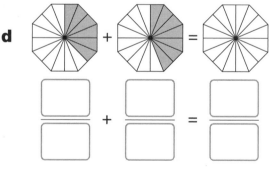

3 Use the diagrams to subtract the fractions. Shade the number of parts shown by the minuend (first fraction) and then cross out the number of shaded parts shown by the subtrahend (second fraction). Then complete the number sentence.

a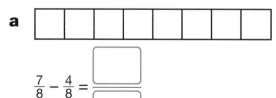

b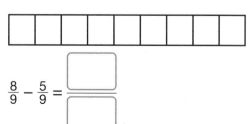

$\frac{7}{8} - \frac{4}{8} = \frac{\square}{\square}$

$\frac{8}{9} - \frac{5}{9} = \frac{\square}{\square}$

c

$\frac{11}{12} - \frac{7}{12} = \boxed{}$

d

$\frac{15}{16} - \frac{8}{16} = \boxed{}$

4 Solve the calculations.

a $\frac{3}{7} + \frac{2}{7} = \boxed{}$

b $\frac{3}{10} + \frac{4}{10} = \boxed{}$

c $\frac{5}{6} + \frac{1}{6} = \boxed{}$

d $\frac{8}{9} - \frac{5}{9} = \boxed{}$

e $\frac{5}{5} - \frac{2}{5} = \boxed{}$

f $\frac{7}{8} - \frac{6}{8} = \boxed{}$

5 Solve the word problems.

a Emilio has \$78. He spends $\frac{3}{6}$ of his money on a model car and $\frac{1}{6}$ of it on a set of stickers for the car.

i What fraction of his original amount of money is left? $\boxed{}$

ii How much money does Emilio have left? \$ $\boxed{}$

b There are 69 learners in Stage 5. $\frac{3}{23}$ of the learners play football, while $\frac{7}{23}$ play netball. No one plays both sports.

i What fraction play football or netball? $\boxed{}$

ii How many more learners play netball than football? Write the answer as a fraction. $\boxed{}$

iii What fraction of learners do not play football or netball? $\boxed{}$

How many learners is that? $\boxed{}$

Date: _____

Lesson 3: **Adding and subtracting fractions (2)**

- Add and subtract fractions with different denominators

1 Use the fraction wall to convert the fractions.

a $\frac{1}{4} = \frac{\boxed{}}{8}$

b $\frac{1}{4} = \frac{\boxed{}}{12}$

c $\frac{1}{2} = \frac{\boxed{}}{6}$

d $\frac{1}{2} = \frac{\boxed{}}{12}$

e $\frac{1}{3} = \frac{\boxed{}}{6}$

f $\frac{1}{3} = \frac{\boxed{}}{9}$

1											
$\frac{1}{2}$						$\frac{1}{2}$					
$\frac{1}{3}$				$\frac{1}{3}$				$\frac{1}{3}$			
$\frac{1}{4}$			$\frac{1}{4}$			$\frac{1}{4}$			$\frac{1}{4}$		
$\frac{1}{5}$		$\frac{1}{5}$		$\frac{1}{5}$		$\frac{1}{5}$		$\frac{1}{5}$			
$\frac{1}{6}$		$\frac{1}{6}$		$\frac{1}{6}$		$\frac{1}{6}$		$\frac{1}{6}$		$\frac{1}{6}$	
$\frac{1}{8}$	$\frac{1}{8}$	$\frac{1}{8}$	$\frac{1}{8}$	$\frac{1}{8}$	$\frac{1}{8}$	$\frac{1}{8}$	$\frac{1}{8}$				
$\frac{1}{9}$	$\frac{1}{9}$	$\frac{1}{9}$	$\frac{1}{9}$	$\frac{1}{9}$	$\frac{1}{9}$	$\frac{1}{9}$	$\frac{1}{9}$	$\frac{1}{9}$			
$\frac{1}{10}$	$\frac{1}{10}$	$\frac{1}{10}$	$\frac{1}{10}$	$\frac{1}{10}$	$\frac{1}{10}$	$\frac{1}{10}$	$\frac{1}{10}$	$\frac{1}{10}$	$\frac{1}{10}$		
$\frac{1}{12}$	$\frac{1}{12}$	$\frac{1}{12}$	$\frac{1}{12}$	$\frac{1}{12}$	$\frac{1}{12}$	$\frac{1}{12}$	$\frac{1}{12}$	$\frac{1}{12}$	$\frac{1}{12}$	$\frac{1}{12}$	$\frac{1}{12}$

2 Convert the fractions.

a $\frac{3}{4} = \frac{\boxed{}}{8}$

b $\frac{3}{6} = \frac{\boxed{}}{12}$

c $\frac{2}{3} = \frac{\boxed{}}{9}$

d $\frac{2}{3} = \frac{\boxed{}}{12}$

e $\frac{2}{5} = \frac{\boxed{}}{10}$

f $\frac{4}{5} = \frac{\boxed{}}{15}$

g $\frac{7}{3} = \frac{\boxed{}}{9}$

h $\frac{5}{4} = \frac{\boxed{}}{8}$

i $\frac{7}{4} = \frac{\boxed{}}{12}$

j $\frac{9}{5} = \frac{\boxed{}}{10}$

k $\frac{8}{7} = \frac{\boxed{}}{21}$

l $\frac{10}{9} = \frac{\boxed{}}{27}$

Number

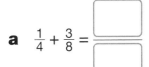 Use the fraction wall to help you add and subtract unlike fractions. Remember to estimate first, e.g. $\frac{7}{8}$ is close to 1 so $\frac{7}{8} - \frac{1}{4}$ should be close to $1 - \frac{1}{4} = \frac{3}{4}$.

a $\frac{1}{4} + \frac{3}{8} = \boxed{}$

b $\frac{7}{8} - \frac{1}{4} = \boxed{}$

c $\frac{3}{5} + \frac{1}{10} = \boxed{}$

d $\frac{4}{5} - \frac{7}{10} = \boxed{}$

e $\frac{1}{12} + \frac{3}{4} = \boxed{}$

f $\frac{11}{12} - \frac{1}{2} = \boxed{}$

g $\frac{3}{4} - \frac{7}{12} = \boxed{}$

h $\frac{11}{10} + \frac{3}{5} = \boxed{}$

i $\frac{9}{10} - \frac{1}{2} = \boxed{}$

j $\frac{9}{5} - \frac{3}{10} = \boxed{}$

k $\frac{19}{12} + \frac{5}{3} = \boxed{}$

l $\frac{11}{12} - \frac{3}{4} = \boxed{}$

4 Solve the problems. Remember to estimate first. In part a the calculation is $\frac{3}{8} + \frac{6}{24}$. $\frac{3}{8}$ is almost $\frac{1}{2}$ and $\frac{6}{24}$ is a $\frac{1}{4}$. So the answer will be near $\frac{3}{4}$.

a $\frac{3}{8}$ of the fish in a tank are blue and $\frac{6}{24}$ of the fish are green. What fraction of the fish in the tank are blue or green?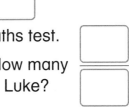

b   Aisha runs $\frac{17}{25}$ of a kilometre. Her friend Chen runs $\frac{2}{5}$ of a kilometre. How much further has Aisha run than Chen? km

c   Luke correctly answers $\frac{5}{6}$ of the questions in a Maths test. His friend Hassan gets $\frac{47}{48}$ of the questions right. How many more questions did Hassan answer correctly than Luke? Write the answer as a fraction.

5 Solve the subtractions.

a $\frac{2}{2} - \frac{1}{4} - \frac{1}{8} = \boxed{}$

b $\frac{2}{3} - \frac{1}{6} - \frac{5}{12} = \boxed{}$

c $\frac{3}{4} - \frac{3}{8} - \frac{3}{16} = \boxed{}$

Date: _____

Number

Lesson 4: **Multiplying and dividing unit fractions**

• Multiply and divide unit fractions

You will need
• coloured pencil

1 Shade the parts to find the sum.

a

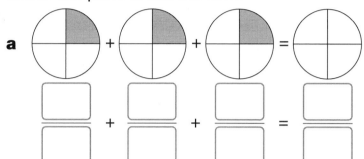

$$\frac{\boxed{}}{\boxed{}} + \frac{\boxed{}}{\boxed{}} + \frac{\boxed{}}{\boxed{}} = \frac{\boxed{}}{\boxed{}}$$

b

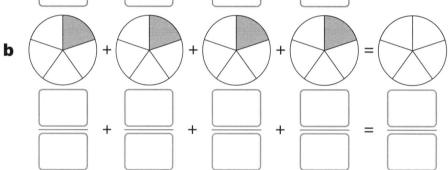

$$\frac{\boxed{}}{\boxed{}} + \frac{\boxed{}}{\boxed{}} + \frac{\boxed{}}{\boxed{}} + \frac{\boxed{}}{\boxed{}} = \frac{\boxed{}}{\boxed{}}$$

2 Use the models to multiply.

a $\frac{1}{10} \times 3 = \dfrac{\boxed{}}{\boxed{}}$

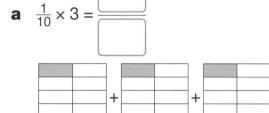

b $\frac{1}{8} \times 4 = \dfrac{\boxed{}}{\boxed{}}$

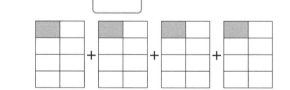

c $\frac{1}{4} \times 5 = \dfrac{\boxed{}}{\boxed{}}$

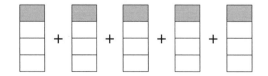

d $\frac{1}{3} \times 7 = \dfrac{\boxed{}}{\boxed{}}$

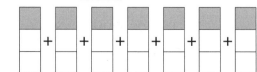

Number

3 Draw models to divide.

$\frac{1}{4} \div 3 =$

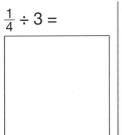

Begin with a whole.

Split the whole into 4 equal parts. Shade 1 part.

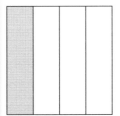

This gives the size of $\frac{1}{4}$.

Next, divide by 3 by splitting each $\frac{1}{4}$ into 3 equal parts.

$\frac{1}{12}$	$\frac{1}{12}$	$\frac{1}{12}$	$\frac{1}{12}$
$\frac{1}{12}$	$\frac{1}{12}$	$\frac{1}{12}$	$\frac{1}{12}$
$\frac{1}{12}$	$\frac{1}{12}$	$\frac{1}{12}$	$\frac{1}{12}$

Each part is $\frac{1}{12}$ of the whole.

So,

$\frac{1}{4} \div 3 = \frac{1}{12}$

a $\frac{1}{3} \div 2 = \frac{\boxed{}}{\boxed{}}$

b $\frac{1}{5} \div 2 = \frac{\boxed{}}{\boxed{}}$

c $\frac{1}{6} \div 3 = \frac{\boxed{}}{\boxed{}}$

d $\frac{1}{5} \div 4 = \frac{\boxed{}}{\boxed{}}$

4 Kisha is given the problem $\frac{1}{2} \div 4$ to solve. She solves it in three steps.

Step 1: She draws a rectangle and divides it into two equal parts.

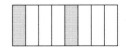

Step 2: Next she divides each half into four equal parts.

Step 3: Then she shades one part out of four and gives the answer $\frac{2}{8}$.

Is Kisha correct? How do you know? _____

Which steps in Kisha's solution would you use? Why? _____

Which steps would you not use? Why? _____

Date: _____

109

Number

Lesson 1: **Percentages of shapes**

• Recognise percentages of shapes

You will need
• coloured pencil

1 The shaded part of each 100 grid represents a percentage. Write the percentage shown.

a
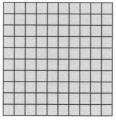

[   ] %

b
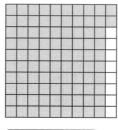

[   ] %

c
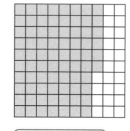

[   ] %

d
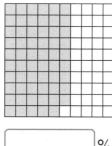

[   ] %

e
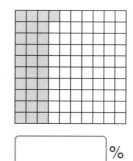

[   ] %

f
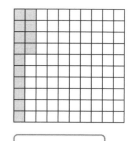

[   ] %

2 Shade each grid to show the percentage given.

a 23%

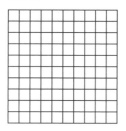

b 34%

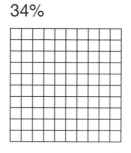

c 51%

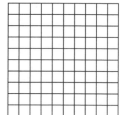

d 67%

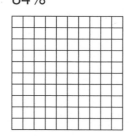

e 84%

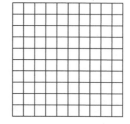

f 96%

Write each statement as a percentage.

a 38 in every 100 T-shirts are white.   **b** 63 in every 100 days are sunny.

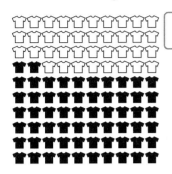 ☐ %

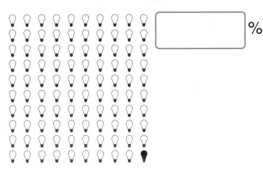 ☐ %

c 77 in every 100 people are happy.   **d** 99 in every 100 lights are on.

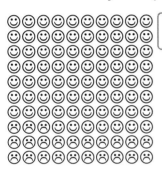 ☐ %

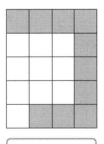 ☐ %

4 Write the percentage of each shape that is shaded.

a

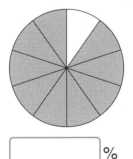

☐ %

b

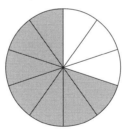

☐ %

c

☐ %

d

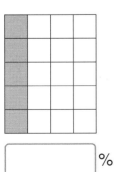

☐ %

e

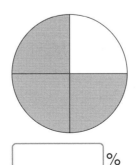

☐ %

f

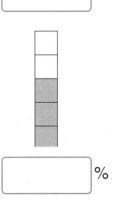

☐ %

Date: _____

Lesson 2: **Converting between fractions and percentages**

Number

• Write percentages as a fraction
with denominator 100

You will need
• coloured pencil

1 Write the fraction, in hundredths, that is represented by the shaded part of each 100 grid. Convert this to a percentage.

a

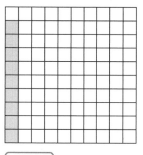

$$\frac{\boxed{}}{\boxed{}} = \boxed{}\%$$

b

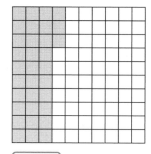

$$\frac{\boxed{}}{\boxed{}} = \boxed{}\%$$

c

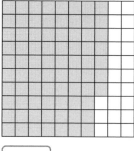

$$\frac{\boxed{}}{\boxed{}} = \boxed{}\%$$

2 Shade the fraction of the grid shown and write the percentage.

a $\frac{1}{2} = \boxed{}\%$

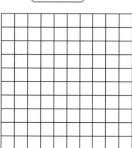

b $\frac{1}{10} = \boxed{}\%$

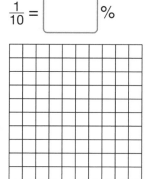

c $\frac{1}{100} = \boxed{}\%$

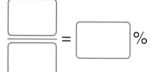

3 Write each statement as a percentage and a fraction.

a 1 out of every 2 balls is green.

$$\boxed{}\% = \frac{\boxed{}}{\boxed{}}$$

b 23 out of every 100 vehicles are motorbikes.

$$\boxed{}\% = \frac{\boxed{}}{\boxed{}}$$

Number

c 1 out of every 10 chocolates in a box is toffee flavoured.

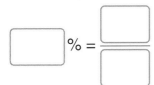 % = ☐/☐

d 61 out of every 100 shapes are triangles.

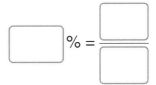 % = ☐/☐

e 1 pair out of every 100 pairs of footwear are sandals.

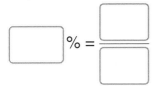 % = ☐/☐

f 93 out of every set of 100 dishes are plastic.

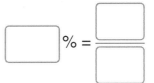 % = ☐/☐

4 Write the equivalent fraction for these percentages.

a 23% = ☐/☐

b 67% = ☐/☐

c 59% = ☐/☐

d 10% = ☐/☐

e 1% = ☐/☐

f 50% = ☐/☐

5 For each fraction, write the percentage equivalent.

a $\frac{1}{2}$ = ☐ %

b $\frac{1}{4}$ = ☐ %

c $\frac{3}{4}$ = ☐ %

d $\frac{7}{10}$ = ☐ %

e $\frac{40}{100}$ = ☐ %

f $\frac{75}{100}$ = ☐ %

6 Leo says: To change a fraction into a percentage, all you need to do is put a per cent sign next to the numerator.

Is Leo's statement always true, sometimes true or never true?

☐ Explain your answer.

_____

_____

_____

Date: _____

Lesson 3: **Comparing percentages**

• Compare percentages and tenths

1 Compare percentages using the symbols <, > or =.

a 50% ☐ 49%   **b** 56% ☐ 65%   **c** 99% ☐ 98%

d 100% ☐ $\frac{100}{100}$ **e** 77% ☐ 79%   **f** 40% ☐ 39%

g 81% ☐ 18%   **h** 4% ☐ 30%   **i** 21% ☐ 12%

2 Mark each pair of numbers on the number line. Draw a ring around the greater number and complete the number statement with the symbol < or >.

a 70% ☐ $\frac{5}{10}$

b $\frac{8}{10}$ ☐ 90%

c 40% ☐ $\frac{3}{10}$

d $\frac{7}{10}$ ☐ 80%

3 Mark each pair of numbers on the number line. Draw a ring around the greater number and complete the number statement with the symbol < or >.

a 70% ☐ $\frac{3}{4}$

Number

b $\frac{1}{4}$ ☐ 20%

0 ——————————————— 1

c 10% ☐ $\frac{1}{5}$

0 ——————————————— 1

d $\frac{3}{5}$ ☐ 70%

0 ——————————————— 1

e 90% ☐ $\frac{4}{5}$

0 ——————————————— 1

f $\frac{2}{5}$ ☐ 50%

0 ——————————————— 1

4 Draw a ring around the greater amount in each pair.

a 10% of 20 cookies   or   20% of 20 cookies

b 50% of $100   or   40% of $100

c 30% of 60 beads   or   25% of 60 beads

d 20% of 40 apples   or   25% of 40 apples

e 98% of 100 people   or   89% of 100 people

f 12% of 100 peanuts   or   21% of 100 peanuts

Date: _____

Lesson 4: **Ordering percentages**

• Order percentages and tenths

1 Write the percentages in order, from smallest to largest.

a 49%   47%   48%   46%   ☐ < ☐ < ☐ < ☐

b 56%   63%   65%   36%   ☐ < ☐ < ☐ < ☐

c 93%   9%   91%   90%   ☐ < ☐ < ☐ < ☐

d 10%   11%   8%   1%   ☐ < ☐ < ☐ < ☐

2 Mark the numbers on the number line. Then write the numbers in order from smallest to greatest.

a 80%, $\frac{5}{10}$, 60%

☐ < ☐ < ☐

b 10%, 30%, $\frac{2}{10}$

☐ < ☐ < ☐

c $\frac{9}{10}$, 70%, 80%

☐ < ☐ < ☐

d 60%, $\frac{7}{10}$, 50%

☐ < ☐ < ☐
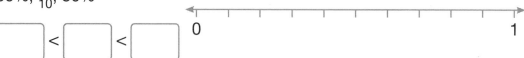

Number

3 Mark the numbers on the number line. Then write the numbers in order from smallest to greatest.

a 70%, 90%, $\frac{3}{4}$

☐ < ☐ < ☐

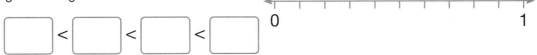

0   1

b $\frac{3}{5}$, 40%, 50%,

☐ < ☐ < ☐

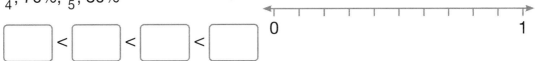

0   1

c $\frac{4}{5}$, 30%, $\frac{2}{5}$, 10%

☐ < ☐ < ☐ < ☐

0   1

d $\frac{1}{5}$, 90%, 50%, $\frac{3}{5}$

☐ < ☐ < ☐ < ☐

0   1

e $\frac{3}{4}$, 70%, $\frac{2}{5}$, 80%

☐ < ☐ < ☐ < ☐

0   1

4 Write the numbers in the order given by the symbols.

a 80%, $\frac{1}{2}$, 40%, $\frac{3}{10}$, 70%

☐ < ☐ < ☐ < ☐ < ☐

b 50%, $\frac{3}{4}$, 20%, $\frac{7}{10}$, 40%

☐ > ☐ > ☐ > ☐ > ☐

c 60%, $\frac{4}{5}$, 40%, $\frac{9}{10}$, 20%

☐ < ☐ < ☐ < ☐ < ☐

d 90%, $\frac{3}{5}$, 10%, $\frac{1}{4}$, 30%

☐ > ☐ > ☐ > ☐ > ☐

Date: _____

Lesson 1: **Adding decimals (1)**

- Add pairs of decimals mentally

1 Add the decimals mentally.

a $0.3 + 0.5 =$ ☐ **b** $0.7 + 0.2 =$ ☐ **c** $0.2 + 0.7 =$ ☐

d $0.4 + 0.4 =$ ☐ **e** $0.22 + 0.14 =$ ☐ **f** $0.45 + 0.23 =$ ☐

g $0.65 + 0.22 =$ ☐ **h** $0.58 + 0.41 =$ ☐ **i** $2.3 + 3.4 =$ ☐

j $4.2 + 2.6 =$ ☐ **k** $3.6 + 5.2 =$ ☐ **l** $6.7 + 3.2 =$ ☐

 2 Add the numbers mentally. Use any strategy you prefer. Remember to estimate first, e.g. $7.8 + 9.9$ will be close to $8 + 10 = 18$.

a $10.3 + 5.9 =$ ☐ **b** $7.8 + 9.9 =$ ☐ **c** $8.7 + 8.7 =$ ☐

d $15.6 + 13.9 =$ ☐ **e** $23.7 + 19.8 =$ ☐ **f** $24.8 + 22.7 =$ ☐

g $33.8 + 32.9 =$ ☐ **h** $41.3 + 16.8 =$ ☐ **i** $0.65 + 0.33 =$ ☐

j $0.78 + 0.54 =$ ☐ **k** $0.83 + 0.69 =$ ☐ **l** $0.77 + 0.94 =$ ☐

m $3.78 + 2.46 =$ ☐ **n** $4.59 + 4.77 =$ ☐ **o** $8.67 + 9.46 =$ ☐

p $3.84 + 4.38 =$ ☐ **q** $5.34 + 7.69 =$ ☐ **r** $2.23 + 5.98 =$ ☐

Describe how you solved questions a, i and n.

a: _____

i: _____

n: _____

 3 Complete the addition pyramids. Each number in the pyramid is the sum of the numbers in the two bricks directly below it. The first row has been done for you.

Number

a

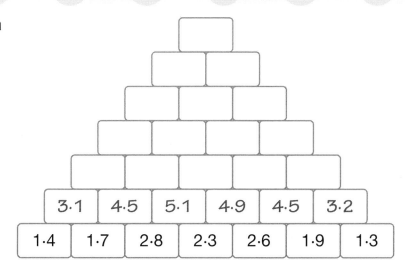

3·1	4·5	5·1	4·9	4·5	3·2

1·4	1·7	2·8	2·3	2·6	1·9	1·3

b

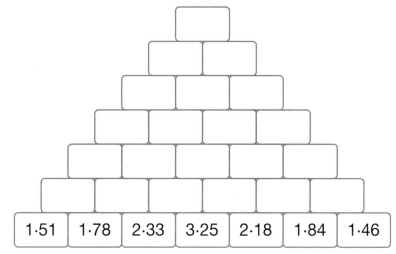

1·51	1·78	2·33	3·25	2·18	1·84	1·46

4 Fill in the missing numbers.

a 0·6 + ☐ = 1·4　　**b** ☐ + 0·7 = 1·1　　**c** 1·3 = 0·5 + ☐

d 0·9 = ☐ + 0·2　　**e** 7·6 = ☐ + 1·2　　**f** ☐ + 4·9 = 11·3

g 11·3 = 6·5 + ☐　　**h** 1·2 + ☐ = 9·7　　**i** 0·3 + ☐ = 0·79

j ☐ + 1·75 = 2·58　**k** 3·3 = 2·88 + ☐　　**l** 5·96 = ☐ + 3·22

m 0·54 + ☐ = 0·79　**n** ☐ + 1·25 = 2·47　**o** 7·2 = 5·7 + ☐

p 8·73 = ☐ + 5·42　**q** 9·45 + ☐ = 12·66　**r** 8·04 = 5·52 + ☐

Date: _____

☺ 😐 ☹

Lesson 2: **Adding decimals (2)**

> **You will need**
> • paper for working out

• Add pairs of decimals using written methods

1 Find two numbers that make the sum. The first one has been done for you.

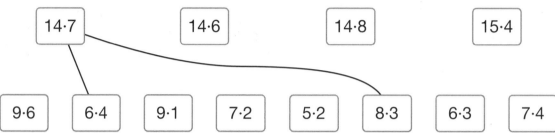

2 **a** Solve by partitioning both numbers. Remember to estimate first, e.g. $7·8 + 8·8$ will be close to $8 + 9 = 17$.

Example: $5·8 + 6·7 = 5 + 6 + 0·8 + 0·7 = 11 + 1·5 = 12·5$

i $2·7 + 3·6 =$

□ + □ + □ + □ = □ + □ = □

ii $6·6 + 1·7 =$

□ + □ + □ + □ = □ + □ = □

iii $8·7 + 7·4 =$

□ + □ + □ + □ = □ + □ = □

iv $7·8 + 8·8 =$

□ + □ + □ + □ = □ + □ = □

b Solve by partitioning one number.

Example: $3·7 + 5·6 = 3·7 + 5 + 0·6 = 8·7 + 0·6 = 9·3$

i $6·5 + 2·7 =$ □ + □ + □ = □ + □ = □

ii $5·8 + 3·4 =$ □ + □ + □ = □ + □ = □

iii $9·6 + 5·6 =$ □ + □ + □ = □ + □ = □

iv $8·7 + 8·5 =$ □ + □ + □ = □ + □ = □

Number

3 Solve each addition using the expanded written method.

a 25·63 + 17·28 = [　　　]

Estimate: [　　　]

```
      |   |   | . |   |   |
  +   |   |   | . |   |   |
  ----+---+---+-.-+---+----
      |   |   | . |   |   |
      |   |   | . |   |   |
      |   |   | . |   |   |
      |   |   | . |   |   |
  ----+---+---+-.-+---+----
      |   |   | . |   |   |
```

b 54·87 + 28·76 = [　　　]

Estimate: [　　　]

```
      |   |   | . |   |   |
  +   |   |   | . |   |   |
  ----+---+---+-.-+---+----
      |   |   | . |   |   |
      |   |   | . |   |   |
      |   |   | . |   |   |
      |   |   | . |   |   |
  ----+---+---+-.-+---+----
      |   |   | . |   |   |
```

4 Solve each addition using the formal written method.

a 34·57 + 28·77 = [　　　]

Estimate: [　　　]

```
      |   |   | . |   |   |
  +   |   |   | . |   |   |
  ----+---+---+-.-+---+----
      |   |   | . |   |   |
```

b 48·86 + 36·48 = [　　　]

Estimate: [　　　]

```
      |   |   | . |   |   |
  +   |   |   | . |   |   |
  ----+---+---+-.-+---+----
      |   |   | . |   |   |
```

5 Using any strategy you prefer, find the total cost of:

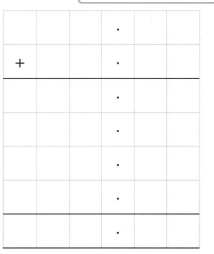

a the tennis racket and the cricket bat $ [　　　]

b the cricket bat and the bike $ [　　　]

c the bike and the wetsuit $ [　　　]

d three items of your choice: [　　　] + [　　　] + [　　　] = [　　　]

$57·66

$68·87

$86·78

$95·96

Date: _____

Lesson 3: **Subtracting decimals (1)**

- Subtract pairs of decimals mentally

1 Subtract the decimals mentally.

a 0·8 – 0·5 = ☐

b 0·7 – 0·2 = ☐

c 0·9 – 0·7 = ☐

d 0·7 – 0·4 = ☐

e 0·36 – 0·14 = ☐

f 0·48 – 0·23 = ☐

g 0·65 – 0·44 = ☐

h 0·98 – 0·41 = ☐

i 4·7 – 1·4 = ☐

j 4·8 – 2·6 = ☐

k 9·6 – 5·2 = ☐

l 8·7 – 4·5 = ☐

 2 Subtract the decimals mentally. Use the box below for jottings. Remember to estimate first, e.g. 20·3 – 10·9 will be close to 20 – 11 = 9.

a 0·43 – 0·23 = ☐

b 0·87 – 0·34 = ☐

c 12·5 – 9·5 = ☐

d 67·3 – 4·2 = ☐

e 20·3 – 10·9 = ☐

f 24·8 – 6·9 = ☐

g 28·6 – 19·8 = ☐

h 25·6 – 17·9 = ☐

i 53·7 – 25·8 = ☐

j 74·2 – 26·7 = ☐

3 Fill in the missing numbers.

a $0.6 - \boxed{} = 0.2$　　**b** $\boxed{} - 0.7 = 0.1$　　**c** $0.3 = 0.8 - \boxed{}$

d $0.6 = \boxed{} - 0.2$　　**e** $7.6 - \boxed{} = 3.2$　　**f** $\boxed{} - 4.9 = 4.3$

g $4.3 = 7.5 - \boxed{}$　　**h** $2.2 = \boxed{} - 3.7$　　**i** $2.76 - \boxed{} = 1.19$

j $\boxed{} - 2.4 = 1.58$　　**k** $5.32 = 7.08 - \boxed{}$　　**l** $3.62 = \boxed{} - 4.82$

4 Each number in the pyramid is the sum of the numbers in the two bricks directly below it. Use the inverse operation of subtraction to fill in the missing numbers in the pyramids. You may still need to use addition to find some of the numbers.

a

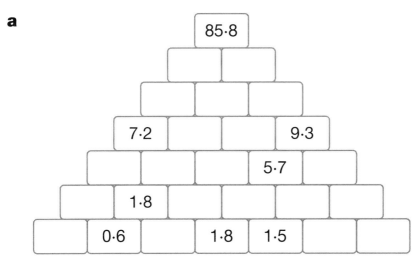

b

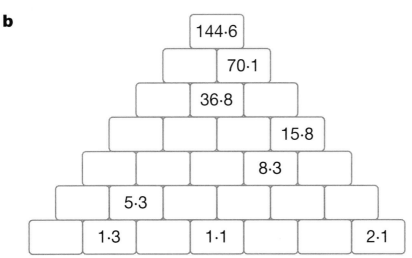

Date: _____

Lesson 4: **Subtracting decimals (2)**

- Subtract pairs of decimals using written methods

You will need
- paper for working out

1 Work out these subtraction calculations. Write your estimate (E) underneath each calculation.

a　　9·6
　　− 5·3
　___

E: [　　　]

b　　87·8
　　− 64·7
　___

E: [　　　]

c　　6·35
　　− 3·23
　___

E: [　　　]

d　　89·67
　　− 36·45
　___

E: [　　　]

e　　8·6
　　− 4·7
　___

E: [　　　]

f　　95·9
　　− 38·7
　___

E: [　　　]

g　　9·47
　　− 7·29
　___

E: [　　　]

h　　73·78
　　− 52·93
　___

E: [　　　]

2 a Solve by partitioning the subtrahend.

Example: 7·4 − 5·7 = 7·4 − 5 − 0·7 = 2·4 − 0·7 = 1·7

i　5·4 − 2·6 = [　] − [　] − [　] = [　] − [　] = [　]

ii　8·3 − 5·8 = [　] − [　] − [　] = [　] − [　] = [　]

iii　9·4 − 7·7 = [　] − [　] − [　] = [　] − [　] = [　]

iv　6·5 − 3·6 = [　] − [　] − [　] = [　] − [　] = [　]

b Use a known number fact to solve each calculation.

Example: 6·3 − 4·7　　63 − 47 = 16 therefore 6·3 − 4·7 = 1·6

i　67 − 38 = [　] therefore [　　　　　　　　]

ii　45 − 18 = [　] therefore [　　　　　　　　]

iii　82 − 36 = [　] therefore [　　　　　　　　]

iv　91 − 67 = [　] therefore [　　　　　　　　]

3 Solve each subtraction using the formal written method.

a   44·55 − 22·38 =

Estimate:

b   54·63 − 27·28 =

Estimate:

c   63·76 − 21·87 =

Estimate:

d   72·63 − 48·85 =

Estimate:

e   84·84 − 56·97 =

Estimate:

f   94·44 − 37·66 =

Estimate:

4 Using any strategy you prefer, find the sale price of each item described.

Sale price

a   Headphones reduced by $17.68 $

b   Radio reduced by $25.76 $

c   Watch reduced by $28.57 $

d   Mobile phone reduced by $36.35 $

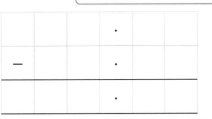
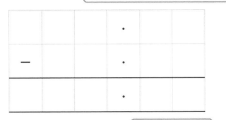
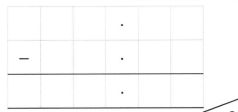
 $42·43
 $63·55
 $94·22
 $86·34

Date: _____

Number

Number

Lesson 1: **Working with place value**

- Use place value to multiply numbers with one decimal place by 1-digit whole numbers

1 Decompose the numbers into tens, ones and tenths.

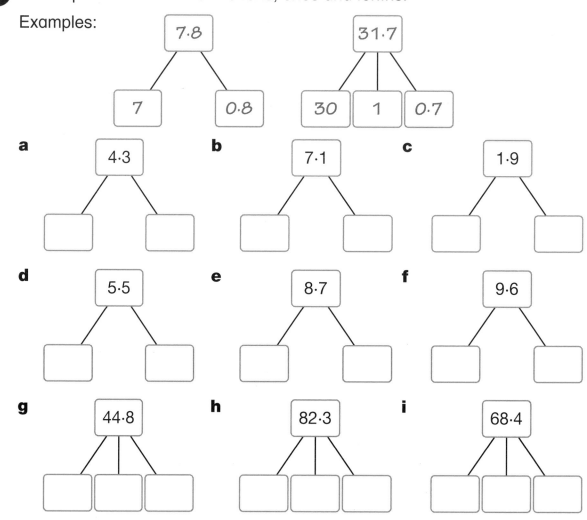

Examples:

7·8 → 7, 0·8

31·7 → 30, 1, 0·7

a 4·3

b 7·1

c 1·9

d 5·5

e 8·7

f 9·6

g 44·8

h 82·3

i 68·4

2 Use the diagram to answer the questions.

Salma has arranged place value counters to show a calculation.

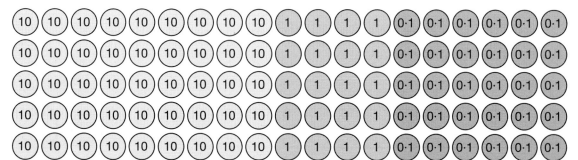

Number

a What is the value of each row?   10s: ☐   1s: ☐ $\frac{1}{10}$s: ☐

b How many rows are there?   10s: ☐   1s: ☐ $\frac{1}{10}$s: ☐

c Write and solve the calculation represented by each group of counters.

10s: ☐ × ☐ = ☐   1s: ☐ × ☐ = ☐

$\frac{1}{10}$s: ☐ × ☐ = ☐

d What calculation is modelled by the counters?   ☐ × ☐ = ☐

 3 Draw a diagram to show each calculation and then use it to find the product. Remember to estimate first, e.g. $21·2 × 4$ will be close to $21 × 4 = 84$.

a $4·4 × 3 =$ ☐

×		
Total		

b $21·2 × 4 =$ ☐

×		
Total		

c $42·3 × 5 =$ ☐

×			
Total			

4 Draw a line to match each calculation to its product.

$84·6 × 3$	$92·7 × 3$	$45·9 × 9$	$51·4 × 8$	$79·3 × 6$	$36·7 × 6$

278·1	475·8	411·2	253·8	220·2	413·1

Date: _____

127

Number

Lesson 2: **Using times table facts**

- Multiply numbers with one decimal place by 1-digit whole numbers, using times table facts and number lines

1 Fill in the missing times table facts.

You will need
- paper for working out

×	1	2	3	4	5	6	7	8	9	10
1	1	2		4	5		7	8	9	10
2	2	4	6		10	12	14	16		20
3	3		9		15		21			30
4	4	8		16		24		32	36	
5	5	10	15	20	25	30	35		45	50
6	6		18				42			
7	7	14		28		42		56	63	70
8	8		24	32	40		56		72	
9	9	18		36		54		72		90
10	10		30		50	60		80	90	

2 Label the number lines in steps of 0·3 and 0·6 to answer the questions.

a

i 0.3 × 3 = ☐　　　**ii** 0.3 × 6 = ☐

iii 0.3 × 4 = ☐　　　**iv** 0.3 × 7 = ☐

b

i 0·6 × 3 = ☐　　　**ii** 0·6 × 6 = ☐

iii 0·6 × 4 = ☐　　　**iv** 0·6 × 7 = ☐

Number

 Find the products. Use your knowledge of times table facts.

a　$0.5 \times 3 =$ 　　　**b**　$0.2 \times 9 =$ 　　　**c**　$0.6 \times 5 =$

d　$0.7 \times 4 =$ 　　　**e**　$0.9 \times 8 =$ 　　　**f**　$0.3 \times 2 =$

g　$0.4 \times 6 =$ 　　　**h**　$0.8 \times 7 =$ 　　　**i**　$0.4 \times 3 =$

j　$0.2 \times 7 =$ 　　　**k**　$0.4 \times 4 =$ 　　　**l**　$0.7 \times 6 =$

4 Solve the word problems.

a　A van contains 8 boxes that have a mass of 0.6 kg each.

What is the total mass of the boxes? 　　　kg

b　Ria's dog eats 0.4 kilograms of dog food each day.

How much food will her dog eat in 7 days? 　　　kg

c　Louis walks 0.8 kilometres each day.

How many kilometres will he walk in 5 days? 　　　km

d　A racing car driver can drive one lap in 0.9 minutes.

How long will it take the driver to drive 7 laps? 　　　minutes

5 Find the outputs. Use any strategy you prefer to multiply the numbers.

a
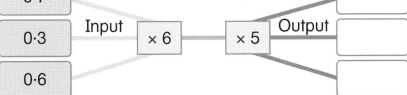

b
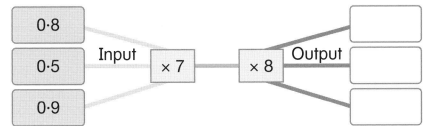

Date: _____

129

Number

Lesson 3: **Grid method**

- Multiply numbers with one decimal place by 1-digit whole numbers, using the grid method

You will need
- paper for working out

1 Use the grid method to multiply.

a $2.6 \times 4 = $ ☐

Estimate: ☐

× ☐ ☐

☐ ☐ ☐

☐

b $4.7 \times 3 = $ ☐

Estimate: ☐

× ☐ ☐

☐ ☐ ☐

☐

c $6.4 \times 5 = $ ☐

Estimate: ☐

× ☐ ☐

☐ ☐ ☐

☐

d $3.5 \times 6 = $ ☐

Estimate: ☐

× ☐ ☐

☐ ☐ ☐

☐

e $7.7 \times 4 = $ ☐

Estimate: ☐

× ☐ ☐

☐ ☐ ☐

☐

f $8.2 \times 8 = $ ☐

Estimate: ☐

× ☐ ☐

☐ ☐ ☐

☐

2 For each calculation, estimate first, then use partitioning to work out the answer. Show your working. Check your answer with your estimate.

a $4.4 \times 6 = $ ☐

Estimate: ☐

b $6.8 \times 7 = $ ☐

Estimate: ☐

Number

c 27·6 × 8 = [ ]

Estimate: [ ]

[ ]

 3 For each calculation, estimate first, then use the grid method to work out the answer. Show your working. Check your answer with your estimate.

a 63·6 × 8 = [ ]

Estimate: [ ]

× [ ] [ ] [ ]

[ ] [ ] [ ] [ ]

[ ]

b 39·4 × 7 = [ ]

Estimate: [ ]

× [ ] [ ] [ ]

[ ] [ ] [ ] [ ]

[ ]

c 58·7 × 6 = [ ]

Estimate: [ ]

× [ ] [ ] [ ]

[ ] [ ] [ ] [ ]

[ ]

d 28·8 × 9 = [ ]

Estimate: [ ]

× [ ] [ ] [ ]

[ ] [ ] [ ] [ ]

[ ]

4 A competition offers five cash prizes. Each prize winner receives an amount paid monthly.

If the amounts are arranged in order, which prize will be in the middle? Draw a ring around it.

PRIZE A	PRIZE B	PRIZE C	PRIZE D	PRIZE E
6 months of $43·70	7 months of $37·80	5 months of $52·20	8 months of $32·60	4 months of $65·90

Date: _____

Lesson 4: **Written methods**

- Multiply numbers with one decimal place by 1-digit whole numbers, using a written method

You will need
- paper for working out

1 Use partitioning to multiply. Estimate the answer first.

a 3·6 × 4 = [ ]

Estimate: [ ]

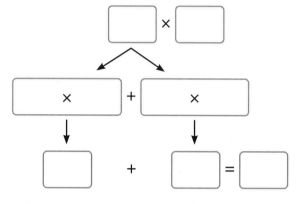

b 6·2 × 5 = [ ]

Estimate: [ ]

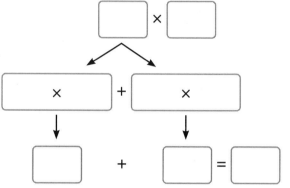

2 Use partitioning to multiply. Estimate the answer first.

a 23·4 × 4 = [ ]

Estimate: [ ]

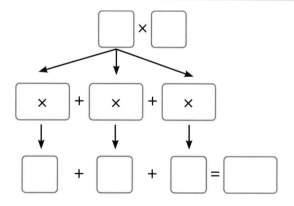

b 35·6 × 5 = [ ]

Estimate: [ ]

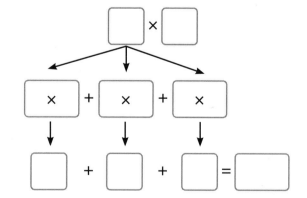

Number

3 Use the expanded written method to multiply. Estimate the answer first.

a $47 \cdot 2 \times 4 =$ [ ] $\times 4 \div 10$

Estimate: [ ]

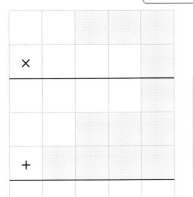

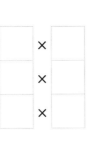

Answer: [ ] $\div 10 =$ [ ]

b $63 \cdot 8 \times 6 =$ [ ] $\times 6 \div 10$

Estimate: [ ]

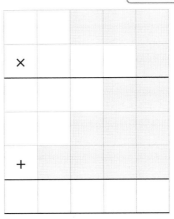

Answer: [ ] $\div 10 =$ [ ]

4 Use any mental or written method to answer these calculations.

a $0 \cdot 7 \times 236 =$ [ ]

b $0 \cdot 3 \times 378 =$ [ ]

c $123 \cdot 4 \times 6 =$ [ ]

d $432 \cdot 6 \times 5 =$ [ ]

Date: _____

Lesson 1: **Fraction, decimal and percentage equivalence (1)**

• Know that proper fractions, decimals and percentages can have equivalent values

You will need
• coloured pencil

1 Write the equivalent fractions.

a $\frac{10}{10} = \dfrac{\boxed{}}{1} = \dfrac{\boxed{}}{2} = \dfrac{\boxed{}}{5} = \dfrac{\boxed{}}{50} = \dfrac{\boxed{}}{100}$

b $\frac{1}{2} = \dfrac{\boxed{}}{4} = \dfrac{\boxed{}}{8} = \dfrac{\boxed{}}{20} = \dfrac{\boxed{}}{80} = \dfrac{\boxed{}}{100}$

2 Shade each fraction on the 100 grid and write the percentage and decimal equivalents.

a $\frac{1}{10} = \boxed{}$ %

$= \boxed{0\cdot}$

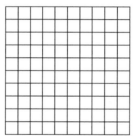

b $\frac{1}{2} = \boxed{}$ %

$= \boxed{0\cdot}$

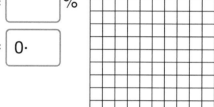

c $\frac{9}{10} = \boxed{}$ %

$= \boxed{0\cdot}$

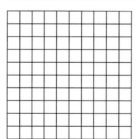

d $\frac{10}{10} = \boxed{}$ %

$= \boxed{\cdot}$

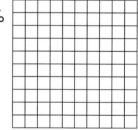

e $\frac{6}{10} = \boxed{}$ %

$= \boxed{0\cdot}$

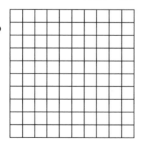

f $\frac{3}{10} = \boxed{}$ %

$= \boxed{0\cdot}$

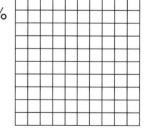

3 Connect each equivalent fraction, percentage and decimal.

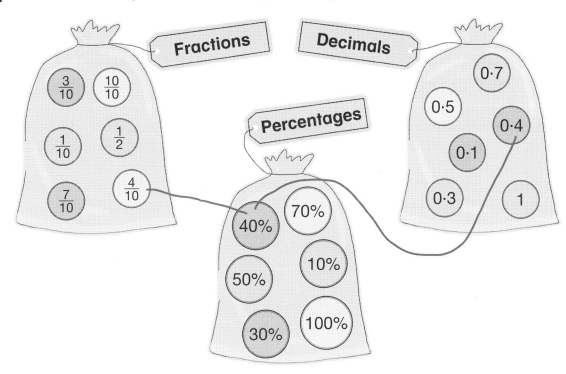

4 Complete the equivalent representations of each number.

a

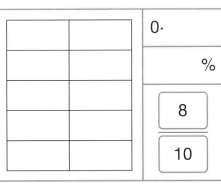

0·6

%

b

0·

20%

c

0·

%

8

10

d

0·

%

Number

Lesson 2: **Fraction, decimal and percentage equivalence (2)**

You will need
- coloured pencil
- paper
- card, ruler, scissors (optional)

- Know that proper fractions, decimals and percentages can have equivalent values

1 a Write each percentage as a decimal.

i 10% = ☐   **ii** 30% = ☐   **iii** 60% = ☐

iv 40% = ☐   **v** 90% = ☐   **vi** 50% = ☐

vii 20% = ☐   **viii** 80% = ☐   **ix** 70% = ☐

b Write each percentage as a fraction.

i 30% = ☐/☐   **ii** 70% = ☐/☐   **iii** 90% = ☐/☐

iv 50% = ☐/☐   **v** 10% = ☐/☐   **vi** 80% = ☐/☐

vii 20% = ☐/☐   **viii** 40% = ☐/☐   **ix** 60% = ☐/☐

2 Shade each fraction on the 100 grid and write the percentage and decimal equivalents.

a $\frac{30}{100}$ = ☐ %   **b** $\frac{90}{100}$ = ☐ %

     = 0·☐   = 0·☐

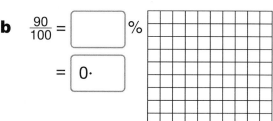

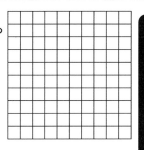

Number

c $\frac{70}{100} = \boxed{}$ %

$= \boxed{0\cdot}$

d $\frac{20}{100} = \boxed{}$ %

$= \boxed{0\cdot}$

3 Complete the equivalent representations of each number.

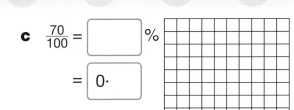

Part of a whole	Part of a group	Fraction	Decimal	Percentage
		$\dfrac{1}{10}$	0·1	10%
		$\dfrac{}{}$	0·2	%
		$\dfrac{}{}$	0·5	%
		$\dfrac{3}{5}$		60%
		$\dfrac{3}{10}$		%
		$\dfrac{}{}$		40%

4 Design a domino set where equivalent percentages, fractions and decimals are matched. If you have time, make the dominoes out of card and play the game.

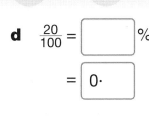

$\frac{30}{100}$ $\frac{7}{10}$ 0·4   0·5

0·7 $\frac{1}{2}$ 30% $\frac{2}{5}$

Date: _____

Number

Lesson 3: Comparing fractions, decimals and percentages

> • Compare numbers with one decimal place, proper fractions and percentages, using the symbols =, > and <

1 a Draw a ring around the greatest number in each row.

| 30% | 60% | 50% |

| 0·4 | 0·3 | 0·5 |

| $\frac{5}{10}$ | $\frac{6}{10}$ | $\frac{4}{10}$ |

b Draw a ring around the smallest number in each row.

| 90% | 60% | 80% |

| 0·9 | 0·8 | 0·7 |

| $\frac{8}{10}$ | $\frac{9}{10}$ | $\frac{10}{10}$ |

2 Use the symbols <, > and = to compare the numbers.

a 20% ⬚ 0·1   **b** $\frac{10}{10}$ ⬚ 100%   **c** $\frac{7}{10}$ ⬚ 0·8

d 60% ⬚ $\frac{1}{2}$ **e** 0·5 ⬚ $\frac{1}{2}$ **f** $\frac{3}{10}$ ⬚ 0·4

g 0·2 ⬚ 20%   **h** 0·6 ⬚ $\frac{7}{10}$ **i** 90% ⬚ $\frac{9}{10}$

j $\frac{8}{10}$ ⬚ 0·8   **k** 4% ⬚ 0·4   **l** 100% ⬚ $\frac{100}{100}$

m 50% ⬚ $\frac{4}{5}$ **n** 5% ⬚ $\frac{1}{2}$ **o** 100% ⬚ 0·1

3 Draw a ring around the greatest number in each row.

30%	$\frac{1}{5}$	$\frac{4}{10}$
$\frac{4}{5}$	95%	0·7
0·5	$\frac{3}{5}$	59%
$\frac{70}{100}$	0·8	74%
85%	0·9	$\frac{4}{5}$
0·3	10%	$\frac{20}{100}$

Number

4 Position each pair of numbers on the number line. Draw a ring around the number that is greater.

a 20% and 0·6

0   1

b $\frac{90}{100}$ and 80%

0   1

c 30% and $\frac{1}{5}$

0   1

d 0·8 and $\frac{3}{5}$

0   1

Date: _____

Number

Lesson 4: **Ordering fractions, decimals and percentages**

- Order numbers with one decimal place, proper fractions and percentages, using the symbols =, > and <

1 Order the fractions.

a $\frac{1}{2}$ $\frac{3}{2}$ $\frac{2}{2}$

☐ < ☐ < ☐

b $\frac{4}{5}$ $\frac{2}{5}$ $\frac{3}{5}$

☐ < ☐ < ☐

c $\frac{8}{10}$ $\frac{7}{10}$ $\frac{3}{10}$

☐ < ☐ < ☐

d $\frac{2}{10}$ $\frac{7}{10}$ $\frac{4}{10}$

☐ < ☐ < ☐

2 Order the decimals.

a 0·4   0·7   0·3

☐ < ☐ < ☐

b 0·2   0·9   0·8

☐ < ☐ < ☐

c 0·7   0·6   0·5

☐ < ☐ < ☐

d 0·5   1·0   0·1

☐ < ☐ < ☐

3 Order the numbers.

a 40%   0·7 $\frac{3}{10}$ 60%   0·2

☐ < ☐ < ☐ < ☐ < ☐

b 0·9 $\frac{1}{2}$ 0·6 $\frac{2}{2}$ 20%

☐ < ☐ < ☐ < ☐ < ☐

c $\frac{7}{10}$ 50%   0·8 $\frac{6}{10}$ 0·4

☐ < ☐ < ☐ < ☐ < ☐

d 0·3   10% $\frac{2}{10}$ 0·5   70%

☐ < ☐ < ☐ < ☐ < ☐

Number

e $\frac{3}{5}$ 0·1   50% $\frac{2}{5}$ 30%

[ ] < [ ] < [ ] < [ ] < [ ]

f 0·5 $\frac{30}{100}$ 40% $\frac{3}{5}$ 35%

[ ] < [ ] < [ ] < [ ] < [ ]

4 Arrange the numbers in descending order.

20% $\frac{4}{5}$ 0·9 $\frac{3}{5}$ 70%   0·5   10%   0·3

[ ] > [ ] > [ ] > [ ] > [ ] > [ ] > [ ] > [ ]

5 Arrange the numbers in ascending order.

a 60% $\frac{4}{5}$ 0·3   0·4 $\frac{1}{5}$

[ ] < [ ] < [ ] < [ ] < [ ]

b 65% $\frac{7}{10}$ 0·6 $\frac{2}{10}$ 0·5

[ ] < [ ] < [ ] < [ ] < [ ]

c 0·1 $\frac{3}{5}$ 15% $\frac{4}{5}$ 0·4

[ ] < [ ] < [ ] < [ ] < [ ]

d 35% $\frac{4}{10}$ 0·2   0·1 $\frac{3}{10}$

[ ] < [ ] < [ ] < [ ] < [ ]

6 Position the numbers on the number line. Then write them in order from smallest to greatest.

a 0·4, $\frac{1}{5}$, 30% and 0·1

[ ] < [ ] < [ ] < [ ] < [ ]

b 70%, $\frac{50}{100}$, 0·6 and $\frac{40}{100}$

[ ] < [ ] < [ ] < [ ] < [ ]

Date: _____

141

Lesson 1: **Proportion (1)**

* Understand that proportion compares part to whole

1 Continue the pattern, then complete the sentence.

a ○△○△○△ ☐ ☐ ☐ ☐ ☐ ☐ ☐ ☐

1 in every ☐ shapes is a triangle.

b ○○○△○○○△ ☐ ☐ ☐ ☐ ☐ ☐ ☐

3 in every ☐ shapes are circles.

c ○○△△△○○△△△ ☐ ☐ ☐ ☐ ☐

2 in every ☐ shapes are circles.

2 Complete the sentences.

a ● ● ○ ● ● ○ ● ● ○

☐ in every ☐ circles is white.

The fraction of circles that are white is ☐/☐

☐ in every ☐ circles are grey.

The fraction of circles that are grey is ☐/☐

b ● ○ ○ ○ ○ ● ○ ○ ○ ○

☐ in every ☐ circles are white.

The fraction of circles that are white is ☐/☐

☐ in every ☐ circles is grey.

The fraction of circles that are grey is ☐/☐

Number

c ◗ ◗ ○ ○ ○ ◗ ◗ ○ ○ ○

☐ in every ☐ circles are white.

The fraction of circles that are white is ☐

☐ in every ☐ circles are grey.

The fraction of circles that are grey is ☐

3 Describe each pattern using the phrase 'in every '. Then write the proportion of white circles as a fraction and the proportion of grey circles as a fraction.

a ◗ ◗ ◗ ◗ ◗ ○ ◗ ◗ ◗ ◗ ◗ ○

_____

_____

_____

b ○ ○ ○ ○ ◗ ◗ ◗ ○ ○ ○ ○ ◗ ◗ ◗

_____

_____

_____

4 Draw a picture to show '2 in every 7'.

5 If 2 out of every 7 pieces of fruit are apples, how many apples are there in a bowl of 28 pieces of fruit? ☐

Date: _____

143

Lesson 2: **Proportion (2)**

- Describe proportions using fractions and percentages

1 Write a sentence to describe each proportion using the phrase '...in every...'.

Example: In every 10 cakes, there is 1 chocolate cupcake.

1 in every 10 cupcakes is chocolate.

a In every 20 kilometres of road there are 4 parking places.

_____

b In every 40 books in a library, there are 8 picture books.

_____

c In every 50 pieces of fruit, there are 20 apples.

_____

d In every 100 runners, there are 55 women.

_____

2 a The children at an after-school club are divided into teams. In every team there are 2 boys and 8 girls. Write the proportion of boys in the team as a fraction and as a percentage. Do the same for the girls.

Boys: ⬜/⬜ ⬜ %   Girls: ⬜/⬜ ⬜ %

b In every 10 chocolates in a box there are 4 dark chocolates and 6 white chocolates. Write the proportion of dark chocolates in the box as a fraction and as a percentage. Do the same for the white chocolates.

Dark: ⬜/⬜ ⬜ %   White: ⬜/⬜ ⬜ %

c Out of every 10 pages in a book there are 3 pages that have pictures. Write the proportion of pages in the book with a picture. Do the same for the pages without pictures.

With: ⬜/⬜ ⬜ %   Without: ⬜/⬜ ⬜ %

3 **a** In every 50 crayons in a tray, there are 25 red crayons and 25 blue crayons.

Write the proportion of red crayons in the tray as a fraction and as a percentage. Do the same for the blue crayons.

Red: ▭/▭ ▭ %   Blue: ▭/▭ ▭ %

b In every 20 hats, 5 are decorated with stars. The rest are decorated with stripes. Write the proportion of hats with stars as a fraction and as a percentage. Do the same for the hats with stripes.

Stars: ▭/▭ ▭ %   Stripes: ▭/▭ ▭ %

c In every 80 pots there are 60 palms. The rest are ferns. Write the proportion of palms in the pots as a fraction and as a percentage. Do the same for the ferns.

Palms: ▭/▭ ▭ %   Ferns: ▭/▭ ▭ %

4 In every 50 ice creams made there are these numbers of each flavour.

Flavour	Number made
mint	5
strawberry	10
cookie	15
bubblegum	20

Write the proportion of each flavour ice cream as a fraction and as a percentage.

Mint: ▭/▭ ▭ %   Strawberry: ▭/▭ ▭ %

Cookie: ▭/▭ ▭ %   Bubblegum: ▭/▭ ▭ %

Date: _____

Number

Lesson 3: **Ratio**

• Understand that ratio compares part to part of two or more quantities

1 Write the ratios.

a ●●○   Black to white: ⬚ to ⬚

b ●○○○ Black to white: ⬚ to ⬚

c In a sports team, there are 4 girls for every 3 boys.

Girls to boys: ⬚ to ⬚

d In a jug of orange squash, there are 2 parts juice for every 3 parts water.

Juice to water: ⬚ to ⬚

2 Write the ratios. Simplify them where possible.

a ●●○○○●●○○○   Black to white: ⬚ to ⬚

b ●○○○○●○○○○   Black to white: ⬚ to ⬚

c An animal charity centre has 20 animals. 15 of the animals are cats

and the rest are dogs. Cats to dogs: ⬚ to ⬚

d A pencil case holds 12 pencils. 8 pencils are red and the rest are blue.

Blue to red pencils: ⬚ to ⬚

3 Express each ratio in two ways.

Example: ●●●○○●●●○○

| Black | to | white | | 3 | : | 2 |
| White | to | black | | 2 | : | 3 |

a ○○●●●●●○○●●●●●

⬚ to ⬚   ⬚ : ⬚

⬚ to ⬚   ⬚ : ⬚

b ○○○○●●●○○○○●●●

[____] to [____]	[__] : [__]
[____] to [____]	[__] : [__]

c ●○○○○◐◐○●○○○○◐◐

[____] to [____] to [____]	[__] : [__] : [__]
[____] to [____] to [____]	[__] : [__] : [__]

d ●●○○○○○○○○◐◐◐●●○○○○○○○○◐◐◐

[____] to [____] to [____]	[__] : [__] : [__]
[____] to [____] to [____]	[__] : [__] : [__]

4 Write the ratios. Simplify them where possible.

a In a bag of red, blue and green cubes the ratio of red cubes to green cubes is 3:5. For every 1 red cube there are two blue cubes.

Write the ratio of red cubes to blue cubes to green cubes.

[__] : [__] : [__]

b In a row of 3D shapes for every 2 cubes there are 4 pyramids. For every 1 cube there are 6 spheres.

Write the ratio of cubes to pyramids to spheres.

[__] : [__] : [__]

c In a cupboard, for every 3 small plates there are 9 large plates. For every 1 medium plate there are 3 large plates.

Write the ratio of small to medium to large plates.

[__] : [__] : [__]

Date: _____

Number

Lesson 4: **Ratio and proportion**

• Use ratio and proportion to make comparisons

1 Convert each fraction to a percentage.

a $\frac{1}{10} = \boxed{}$ %   **b** $\frac{3}{10} = \boxed{}$ %   **c** $\frac{7}{10} = \boxed{}$ %

d $\frac{9}{10} = \boxed{}$ %   **e** $\frac{5}{10} = \boxed{}$ %   **f** $\frac{1}{5} = \boxed{}$ %

g $\frac{3}{5} = \boxed{}$ %   **h** $\frac{2}{5} = \boxed{}$ %   **i** $\frac{4}{5} = \boxed{}$ %

2 Write the ratio and proportion of **black to white** circles.

a ●●●○○○○○○○   Ratio $\boxed{}$: $\boxed{}$

Proportion of black circles   Proportion of white circles

$\frac{\boxed{}}{\boxed{}}$ or $\boxed{}$ % $\frac{\boxed{}}{\boxed{}}$ or $\boxed{}$ %

b ●●●●●○○○○○   Ratio $\boxed{}$: $\boxed{}$

Proportion of black circles   Proportion of white circles

$\frac{\boxed{}}{\boxed{}}$ or $\boxed{}$ % $\frac{\boxed{}}{\boxed{}}$ or $\boxed{}$ %

c ●●○○○○○○○○   Ratio $\boxed{}$: $\boxed{}$

Proportion of black circles   Proportion of white circles

$\frac{\boxed{}}{\boxed{}}$ or $\boxed{}$ % $\frac{\boxed{}}{\boxed{}}$ or $\boxed{}$ %

3 Write the ratio and proportion of **black to white to grey** circles.

a ●○○○○○○○◉◉   Ratio $\boxed{}$: $\boxed{}$: $\boxed{}$

Proportion of:

black circles   white circles   grey circles

$\frac{\boxed{}}{\boxed{}}$ or $\boxed{}$ % $\frac{\boxed{}}{\boxed{}}$ or $\boxed{}$ % $\frac{\boxed{}}{\boxed{}}$ or $\boxed{}$ %

Number

b ●●●○○●●●● Ratio [ ] : [ ] : [ ]

Proportion of:

black circles   white circles   grey circles

[ ]/[ ] or [ ] %   [ ]/[ ] or [ ] %   [ ]/[ ] or [ ] %

c ●●●●●●●○○●● Ratio [ ] : [ ] : [ ]

Proportion of:

black circles   white circles   grey circles

[ ]/[ ] or [ ] %   [ ]/[ ] or [ ] %   [ ]/[ ] or [ ] %

4 Write the ratio and proportion of **pineapples to oranges**. Simplify the ratio where possible.

a 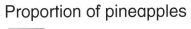   Ratio [ ] : [ ]

Proportion of pineapples   Proportion of oranges

[ ]/[ ] or [ ] %   [ ]/[ ] or [ ] %

b 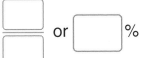   Ratio [ ] : [ ]

Proportion of pineapples   Proportion of oranges

[ ]/[ ] or [ ] %   [ ]/[ ] or [ ] %

c   Ratio [ ] : [ ]

Proportion of pineapples   Proportion of oranges

[ ]/[ ] or [ ] %   [ ]/[ ] or [ ] %

Date: _____

Lesson 1: **Calculating time intervals (1)**

- Understand time intervals of less than one second

1 Compare the lengths of time. Use the symbol < or >.

a	1·7 s	☐	1·6 s	**b**	0·8 s ☐ 0·9 s	
c	2·4 s	☐	2·2 s	**d**	0·3 s ☐ 0·4 s	
e	4·45 s	☐	4·54 s	**f**	7·21 s ☐ 7·12 s	
g	0·98 s	☐	0·99 s	**h**	0·04 s ☐ 0·03 s	
i	0·2 s	☐	0·09 s	**j**	1·67 s ☐ 1·6 s	
k	0·33 s	☐	0·3 s	**l**	8·4 s ☐ 8·39 s	
m	0·76 s	☐	0·6 s	**n**	5·81 s ☐ 5·1 s	
o	9·11 s	☐	9·09 s	**p**	0·09 s ☐ 0·1 s	

2 Order the time spans, from longest to shortest.

a 0·77 s; 0·98 s; 0·54 s; 0·8 s; 0·9 s

b 0·81 s; 0·9 s; 0·15 s; 0·47 s; 0·8 s

c 0·8 s; 0·27 s; 0·7 s; 0·5 s; 0·78 s

d 0·2 s; 0·4 s; 0·7 s; 0·99 s; 0·19 s

e 0·34 s; 0·54 s; 0·5 s; 0·90 s; 0·51 s

f 0·32 s; 0·3 s; 0·10 s; 0·92 s; 0·09 s

g 0·5 s; 0·4 s; 0·05 s; 0·6 s; 0·51 s

h 0·7 s; 0·66 s; 0·8 s; 0·39 s; 0·78 s

Geometry and Measure

3 Draw a line to match each activity to its approximate time span on the number line.

Running 10 metres	Opening a packet of crisps	Writing your name	Folding a square piece of paper in quarters	Zipping up your jacket	Knocking on the door three times

0　　　　1　　　　2　　　　3　　　　4　　　　5

time (s)

 4 Hassan makes a statement about decimal time intervals:

> Decimal times with more digits are always greater amounts of time than decimals with fewer digits.

Is Hassan correct? ☐ Why?

a Explain your answer.

_____

_____

b What would you say or do to change Hassan's thinking?

_____

_____

5 Arrange the letter codes for the time intervals in order, from shortest to longest.

A 53 seconds to 1 min 6 seconds

B 2 min 8 seconds to 3 min 2 seconds

C 1 min 46 seconds to 2 min 42 seconds

D 7 min 56 seconds to 8 min 39 seconds

☐ < ☐ < ☐ < ☐

Date: _____

Geometry and Measure

Lesson 2: **Calculating time intervals (2)**

• Find time intervals in seconds, minutes and hours

Each pair of clocks shows the time four learners started and finished their homework. Calculate the amount of time each learner has spent completing their homework. All the clocks show times in the evening.

1 **a** Leo

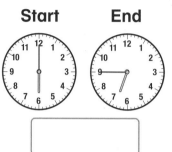

b Rishi

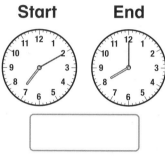

c Milla

d Ria

2 **a** Daniel

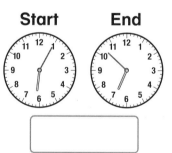

b Manjeet

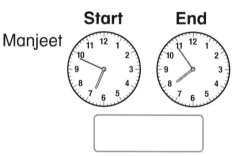

c Leo

d Freya

Geometry and Measure

Calculate the amount of time each person has spent at work.

Name	Start time	Finish time	Time spent at work (hours, min)	Time (min)
Mr Levi	08:15	17:25		
Mrs Taylor	07:35	18:20		
Mr Shah	07:44	18:12		
Mrs Chang	08:23	17:56		

Who spent the longest time at work?

4 Solve these time problems.

a Mrs Davis finished cleaning her house at 17:43. She began cleaning at 14:02. How long did it take her to clean the whole house?

b Mr Ahmed finished gardening at 19:12. He began working in his garden at 11:33. How long did the work take him?

c Pavel arrived home after work at 18:57. He left the house at 07:19 in the morning. How long was he out of the house?

5 The table shows the time it took people to walk a long distance route. Complete the start times, using the 24-hour clock.

Name	Start time	Finish time	Time taken
Clara		12:25	3 hours 13 minutes
Freddy		13:17	4 hours 26 minutes
Jay		13:49	5 hours 57 minutes
Antonio		12:09	4 hours 33 minutes

Date: _____

Geometry and Measure

Lesson 3: **Expressing time intervals**

- Express time intervals as decimals, or in mixed units

1 Complete the time facts.

a 1 min = ☐ seconds **b** 1 hour = ☐ minutes

c 1 day = ☐ hours **d** 1 week = ☐ days

e 1 year = ☐ months **f** 1 year = ☐ weeks

g 1 year = ☐ days or ☐ days

h 1 decade = ☐ years **i** 1 century = ☐ years

2 Complete each table.

a

Minutes	Seconds
2	
6	
9·5	

b

Hours	Minutes
3	
7	
10·5	

c

Days	Hours
2	
5	
8	

d

Weeks	Days
13	
17	
26	

e

Years	Months
8	
18	
29	

f

Years	Weeks
5	
14	
27	

3 Convert these decimal times.

a 0·5 h = ☐ min **b** 0·25 h = ☐ min **c** 0·75 h = ☐ min

d 0·5 min = ☐ s **e** 0·25 min = ☐ s **f** 0·75 min = ☐ s

g 1·1 min = ☐ s **h** 1·2 min = ☐ s **i** 1·4 min = ☐ s

j 1·5 days = ☐ h **k** 1·25 days = ☐ h **l** 1·75 days = ☐ h

Geometry and Measure

4 Lucy and Hassan each complete an art project over four evenings.
The amount of time spent working on the project is given in the table below.

	Monday	Tuesday	Wednesday	Thursday
Lucy	1·5 h	45 min	1·2 h	1 hr 15 min
Hassan	70 min	1 hr 20 min	1·25 h	55 min

a Convince them that Lucy has spent more time on her art project.

_____

b How much more time did Lucy spend on her project than Hassan?

_____

5 Work out the answers to these problems. Show your working.

a Mrs Boucher spends 18 000 seconds a day at work. She has
45 minutes for lunch. How many hours does she work in total? ☐ h

b The Khan family go on holiday and are away for 216 hours.
How many more hours must they be away to make 2 weeks? ☐ h

c A building project took 672 days to complete. The builders on the
project worked for 7 hours a day. How many hours did it take to
complete the project? How many days is that? How many weeks?

☐ h   ☐ d   ☐ weeks

Date: _____

Geometry and Measure

Lesson 4: **Compare times between time zones**

Geometry and Measure

- Compare times between different time zones

You will need
- time zones map

1 Use the world time zones map to work out the answers to these questions.

a How many hours must you turn back your watch if you travel from:

i Istanbul to Berlin? ☐ h

ii Beijing to Madrid? ☐ h

iii Tokyo to Stockholm? ☐ h

iv Auckland to Nairobi? ☐ h

b How many hours must you put forward your watch if you travel from:

i Mexico City to Rio de Janeiro? ☐ h

ii Vancouver to Istanbul? ☐ h

iii Lima to Beijing? ☐ h

iv Hawaii to Tokyo? ☐ h

2 Use the world time zones map to work out the times in different cities around the world.

a It is 09:25 in London.   What time is it in Paris? ☐

What time is it in New York? ☐

b It is 11:33 in Moscow.   What time is it in Oslo? ☐

What time is it in Hong Kong? ☐

c It is 14:07 in New York.   What time is it in Los Angeles? ☐

What time is it in Rome? ☐

Geometry and Measure

d It is 16:52 in Istanbul.   What time is it in London?

What time is it in Sydney?

e It is 19:44 in Tokyo.   What time is it in Los Angeles?

What time is it in Auckland?

f It is 22:17 in Beijing.   What time is it in New York?

What time is it in Sydney?

3 Use the world time zones map to answer the questions.

a A TV news programme in Berlin starts at 8:00 a.m.

What time is it in Sydney when the TV bulletin starts?

b Grigory lives in Moscow. He makes a phone call at 9:00 a.m. to his cousin Tatyana, who lives in Los Angeles.

What time does Tatyana receive the call?

4 Bradley flew straight from London to Sydney. He left London at 11:15 Monday (London time) and arrived in Sydney at 18:35 Tuesday (Sydney time). How long was the flight if Sydney is 10 hours ahead of London?

Date: _____

157

Lesson 1: **Identifying, describing and classifying triangles**

Geometry and Measure

- Identify, describe and classify triangles

You will need
- red, green and blue coloured pencils
- ruler

1 Use the key to colour each triangle.

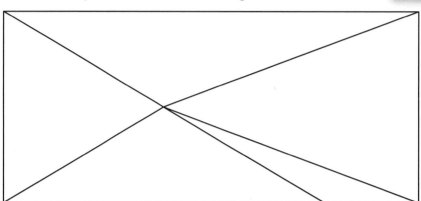

Key

Red – equilateral

Green – scalene

Blue – isosceles

2 Complete the table by classifying, and writing the letter of each shape in the correct column.

Equilateral triangles	Isosceles triangles	Scalene triangles

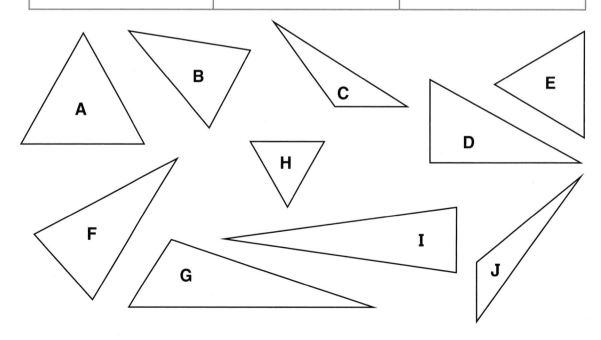

3 Complete the table. Write ✓ for yes and ✗ for no.

Triangle	Equilateral	Isosceles	Scalene
all sides equal, all angles equal			
two sides equal, two angles equal			
no sides equal, no angles equal			
can have a right angle			
can have an angle greater than 90°			
has two or more angles less than 90°			
regular shape			
irregular shape			

Geometry and Measure

4 Divide each shape by drawing a diagonal. Name the triangles it makes. The first one has been done for you.

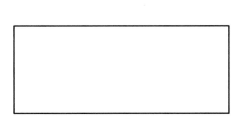

scalene

scalene

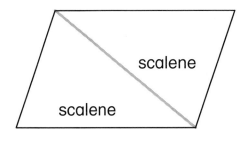

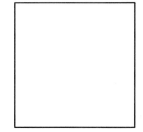

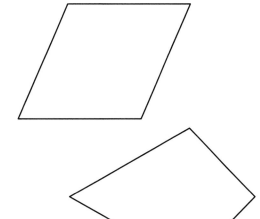

Date: _____

 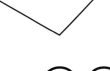

Lesson 2: **Sketching triangles**

• Describe and draw the three types of triangle

You will need
• ruler

Geometry and Measure

1 Use the triangular dot paper to sketch three scalene triangles.

2 Use the triangular dot paper to sketch:

a an isosceles triangle with a base of 5 cm

b an equilateral triangle with sides of 6 cm.

Geometry and Measure

3 Sketch an equilateral triangle with base AB.

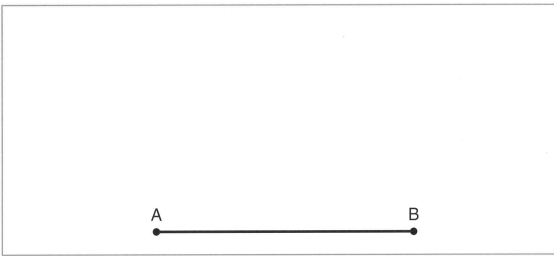

4 Sketch an isosceles triangle ABC with base AB.

5 Billy drew an isosceles triangle but, unfortunately, sides AB and AC were erased. Redraw sides AB and AC for him.

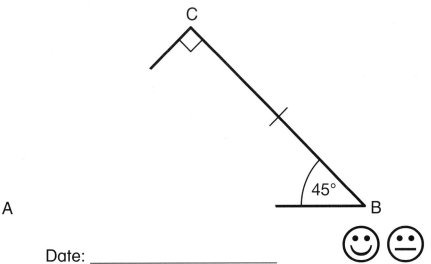

Date: _____

Lesson 3: **Identifying symmetrical patterns**

Geometry and Measure

• Identify symmetrical patterns

You will need
• ruler

1 Draw a ring around the patterns that are symmetrical.

a 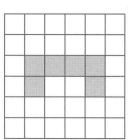 b 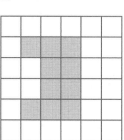 c 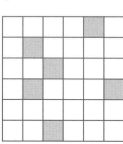 d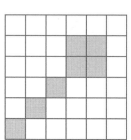

2 Draw the lines of symmetry.

a 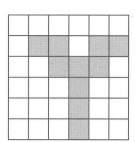 b 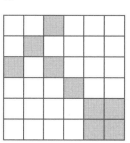 c 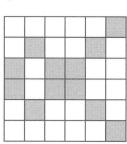 d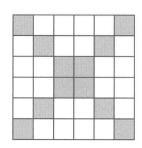

3 Draw a ring around the patterns with reflective symmetry.

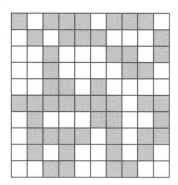 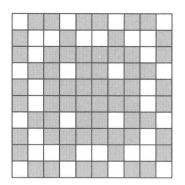 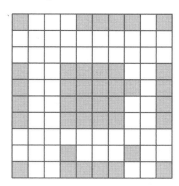

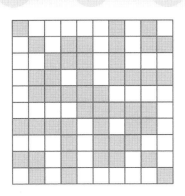
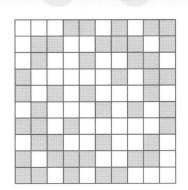
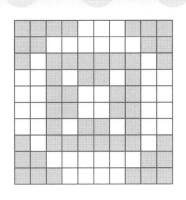

4 Shade each grid to make patterns with 1 or 2 lines of symmetry.

a　　　　　　　　**b**　　　　　　　　**c**

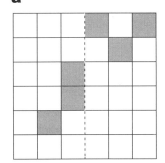
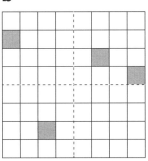
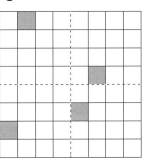

5 Shade in more squares to make the figure symmetrical about a diagonal. Is there more than one way to do it? Will the figure be symmetrical about both diagonals?

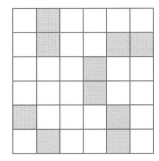

_____

Solution 1　　　　Solution 2　　　　Solution 3

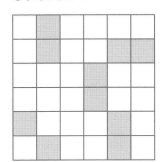
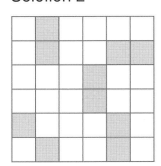
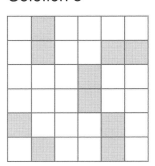

Date: _____

Lesson 4: **Completing symmetrical patterns**

> • Create symmetrical patterns

1 Complete these patterns so that they have one line of symmetry.

a   b   c

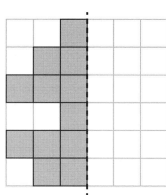

2 Complete these patterns so that they have one line of symmetry.

a   b   c

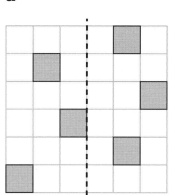 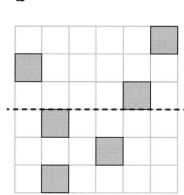 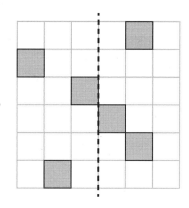

3 Complete these patterns so that they have two lines of symmetry.

a   b   c

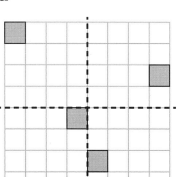 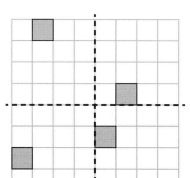 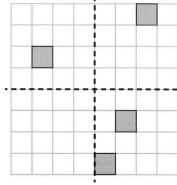

Create your own symmetrical patterns with the pegboards below.

Example:

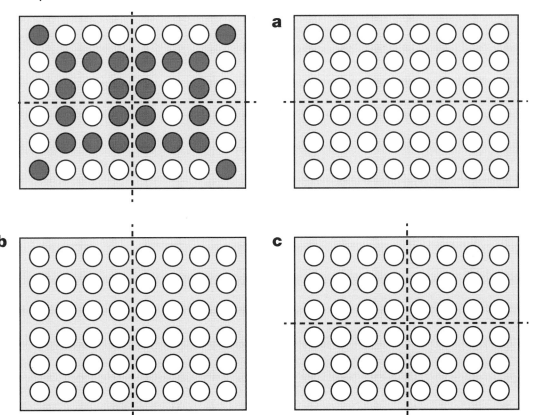

5 Complete the grid designs to make them fully symmetrical.

Use the empty grid to make up your own design.

a   **b**   **c**

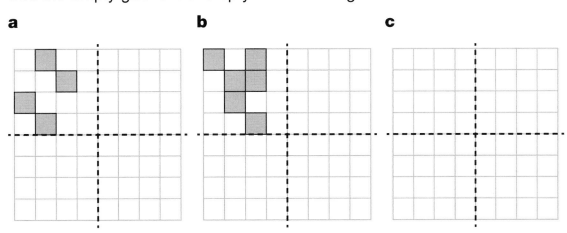

Date: _____

Geometry and Measure

Lesson 1: **Identifying and describing 3D shapes**

> • Identify and describe 3D shapes

1 Name these shapes.

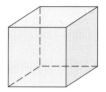   

_____   _____   _____   _____

2 Tick the polyhedra.

 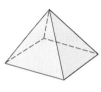   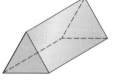

3 Paz combined two cubes together. What new 3D shape did he make?

_____

4 The shape on the right is a face from a 3D shape.

Which 3D shape could it be?

_____

5 Why is a cone not a polyhedron?

_____

6 Name a 3D shape that has:

a no edges   _____

b a rectangle as one of the faces   _____

c at least one flat circular surface _____

d more than 8 faces   _____

Geometry and Measure

7

5

Characterise and list the properties of these shapes. Don't forget to include whether or not they are prisms or pyramids.

a Square-based pyramid

_____

_____

_____

_____

b Tetrahedron

_____

_____

_____

_____

c Hexagonal prism

_____

_____

_____

_____

d Cylinder

_____

_____

_____

_____

8 I have 12 edges, 8 vertices and 6 faces. 4 of the faces are rectangular.

I am a prism. What shape am I? _____

Make up your own clues like this for a 3D shape of your choice.

_____

_____

_____

Date: _____

Lesson 2: **Sketching 3D shapes**

- Describe and sketch 3D shapes

1 Name the 3D shapes.

a

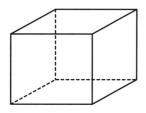

_____

b

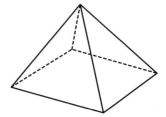

_____

c

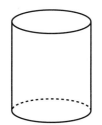

_____

d

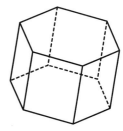

_____

2 Complete these sketches of a cube and a cuboid.

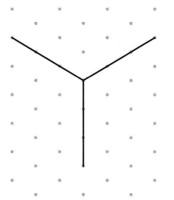 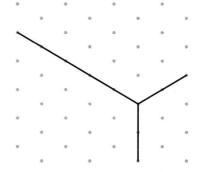

Geometry and Measure

3 Sketch the 3D shape named in each box.

a	b
cube	cuboid
c	**d**
cylinder	cone

4 Draw dashed lines to show the edges and faces of the shapes that are hidden from view. The first one has been done for you.

a 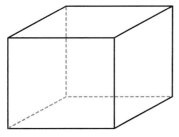   **b** 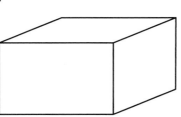   **c**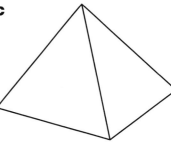

d 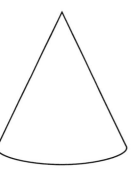   **e** 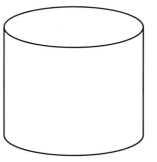   **f**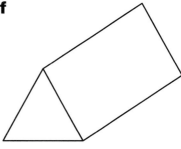

Date: _____

Geometry and Measure

Lesson 3: **Nets for a cube (1)**

- Identify and sketch different nets for a cube

You will need
- squared paper
- ruler
- scissors
- coloured pencil

1 Draw a ring around the correct net for each cube.

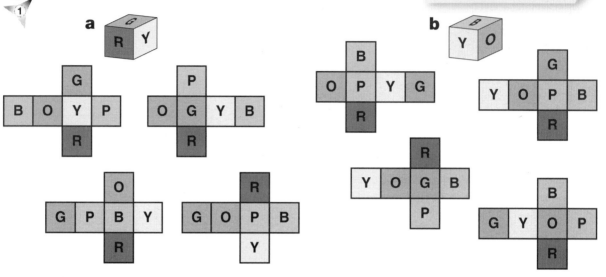

2 Some of the shapes below are nets of open or closed cubes. Tick ✓ all the shapes that are a net of an open or closed cube. If you are not sure, copy the shape onto squared paper, cut it out and try to fold it into a cube to check.

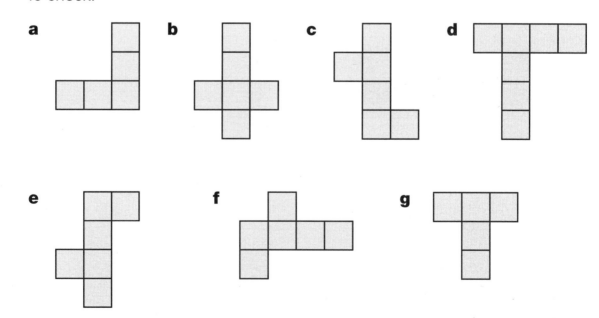

a   b   c   d

e   f   g

Geometry and Measure

3 Look at the three views of each closed cube. Work out where the faces are in relation to each other and colour in the faces of the net.

Views of cube	Net of cube
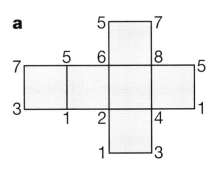	
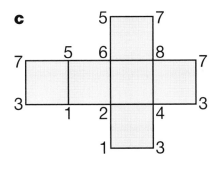	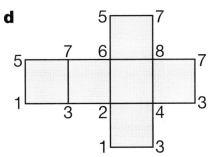

B – Blue
O – Orange
G – Green
Y – Yellow
P – Purple
R – Red

4 Draw a ring around the net that has the vertices of the cube correctly labelled.

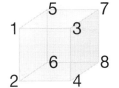

a

```
        5⌐7
   5  6│  8
7 ┌──┬──┤   5
3 └──┴──┤   1
   1  2│  4
       1└─3
```

b

```
          5⌐7
     7  6│  8
5 ┌──┬──┤   5
1 └──┴──┤   1
     3  2│  4
         1└─3
```

c

```
        5⌐7
   5  6│  8
7 ┌──┬──┼──┐ 7
3 └──┴──┼──┘ 3
   1  2│  4
       1└─3
```

d

```
          5⌐7
     7  6│  8
5 ┌──┬──┼──┐ 7
1 └──┴──┼──┘ 3
     3  2│  4
         1└─3
```

Date: _____

171

Lesson 4: **Nets for a cube (2)**

Geometry and Measure

- Identify and sketch different nets for a cube

1 Visualise the folded net in your mind.
Will it form a closed cube?

Draw a ring around your answer.

Yes   No

Explain your answer.

_____

_____

_____

_____

_____

_____

2 Tick all the nets that will build a closed cube and will not have any faces that overlap.

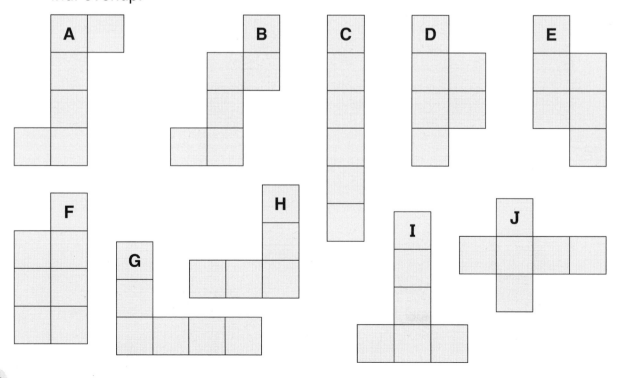

Geometry and Measure

 Sketch a net of a closed cube that has not been included in .

4 Draw a ring around any of nets B, C, D and E that can be folded to produce the same shaded cube as net A.

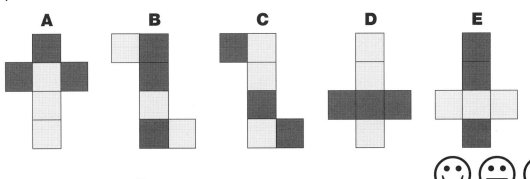

Date: _____

Lesson 1: **Estimating and classifying angles**

• Estimate and classify angles

> **You will need**
> • ruler

1 Write the size of each angle in degrees.
Remember to include the degrees symbol °.

a [ ]

b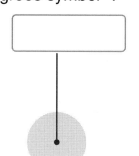

c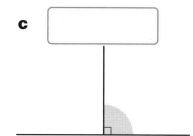

2 Write a letter inside each shaded angle.

A = acute angle

R = right angle

O = obtuse angle

3 Identify and label the angles acute, right, obtuse, straight or reflex.

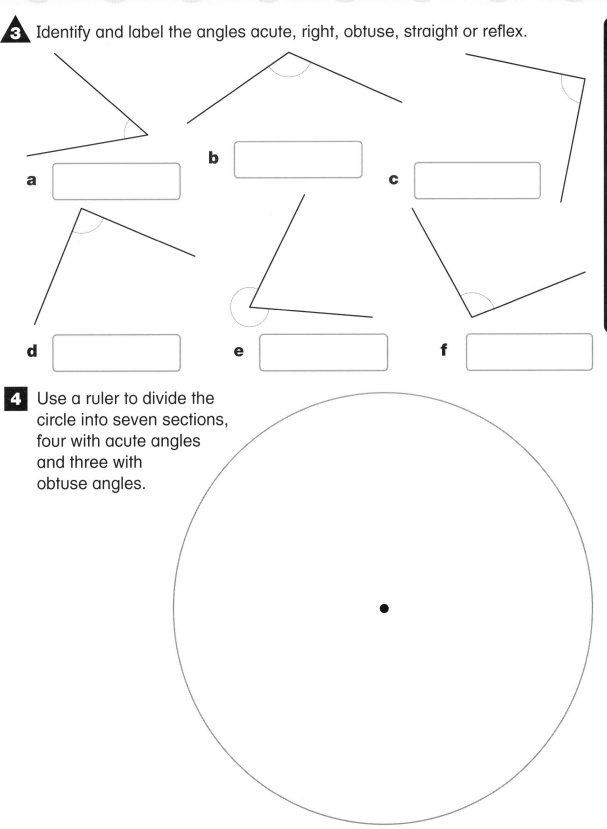

a

b

c

d

e

f

4 Use a ruler to divide the circle into seven sections, four with acute angles and three with obtuse angles.

Date: _____

Lesson 2: **Comparing angles**

Geometry and Measure

- Compare angles

You will need
- tracing paper

1 A, B and C are slices of cake. Identify the angles as acute, right or obtuse. Then write the letter codes of the angles in ascending order.

a

Angle A: _____

b

Angle B: _____

c

Angle C: _____

Order: [ ] < [ ] < [ ]

2 Each angle has a letter. Place the angles in ascending and descending order by writing the letters in the boxes below.

A

B

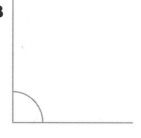

C

D

E

Ascending order: [ ] < [ ] < [ ] < [ ] < [ ]

Descending order: [ ] > [ ] > [ ] > [ ] > [ ]

Geometry and Measure

3 You will need some tracing paper. Trace the angles and then order them, from smallest to largest.

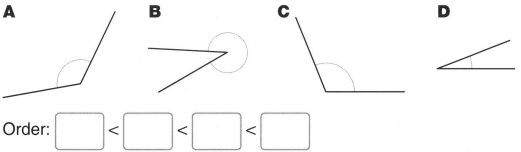

A   **B**   **C**   **D**

Order: ☐ < ☐ < ☐ < ☐

4 Look at the metro map.

⁶ **a** Identify ten angles (including reflex angles) and label the angles A to J. Then sort the angles onto the table by writing their letter codes.

acute	right	obtuse	reflex

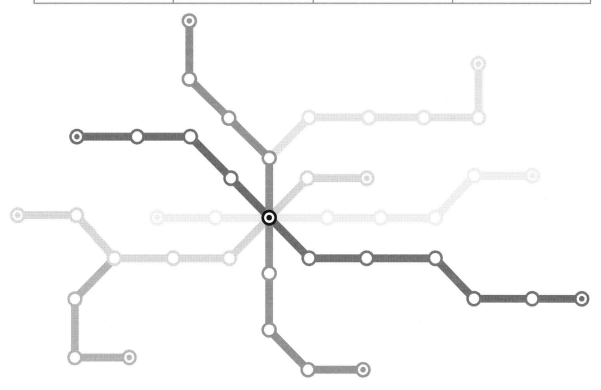

b Choose five angles and write their letter codes in order of size, largest to smallest.

☐ > ☐ > ☐ > ☐ > ☐

Date: _____

Lesson 3: **Angles on a straight line (1)**

- Calculate unknown angles on a straight line

1 Complete each calculation.

a   90 + 90 = ☐

b   100 + ☐ = 180

c   ☐ + 160 = 180

d   110 + 70 = ☐

e   ☐ + 130 = 180

f   150 + ☐ = 180

g   155 + ☐ = 180

h   ☐ + 35 = 180

i   95 + 85 = ☐

j   156 + ☐ = 180

k   ☐ + 72 = 180

l   ☐ + 63 = 180

 2   Write the unknown angles in the boxes. Remember to include the degrees symbol °.

a

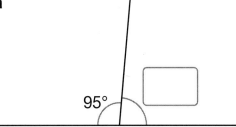

95°

b

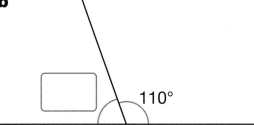

110°

c

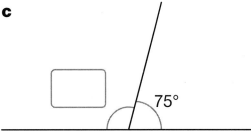

75°

d

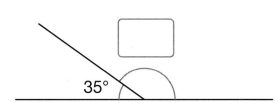

35°

3 Write the unknown angles in the boxes. Remember to include the degrees symbol °.

a

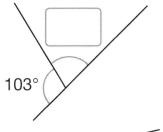

103°

b

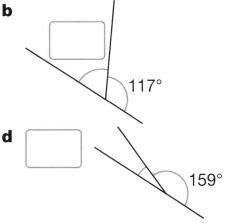

117°

c

59°

d

159°

4 Write the unknown angles in the boxes.

a

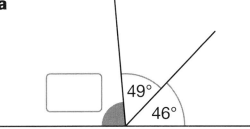

49°
46°

b

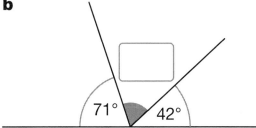

71°
42°

c

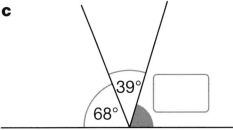

39°
68°

d

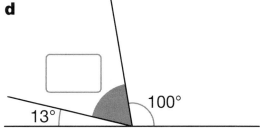

13°
100°

5 Tammy and Leo have half a pizza to share between them.
Tammy would like three times as much as Leo.
The pizza is divided by cutting from the centre
of the straight edge. At which angle should
the pizza be cut?

Date: _____

Lesson 4: **Angles on a straight line (2)**

- Calculate unknown angles on a straight line

1 Complete each calculation.

a 30 + 40 + 110 = ☐

b 100 + 20 + ☐ = 180

c ☐ + 40 + 70 = 180

d 50 + 50 + 80 = ☐

e ☐ + 70 + 50 = 180

f 140 + 10 + ☐ = 180

g 135 + 25 + ☐ = 180

h ☐ + 65 + 35 = 180

i 95 + 45 + 40 = ☐

j 126 + 26 + ☐ = 180

k ☐ + 32 + 52 = 180

l ☐ + 53 + 63 = 180

2 Write the unknown angles in the boxes. Remember to include the degrees symbol °.

a

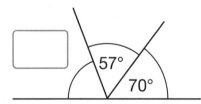

b

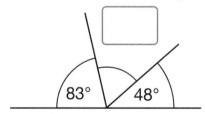

c

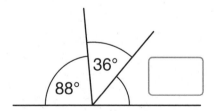

d

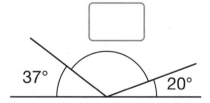

Geometry and Measure

3 Write the unknown angles in the boxes. Remember to include the degrees symbol °.

a
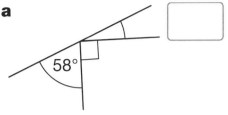
58°

b
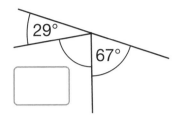
29°
67°

c
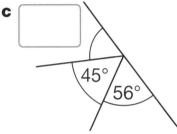
45°
56°

d
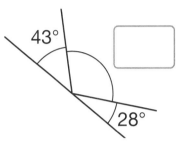
43°
28°

4 Write the value of *b*, *c*, *d* and *e*.

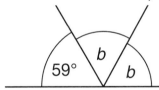
59° *b* *b*

$b = \boxed{}°$

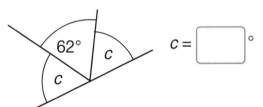
62° *c*
c

$c = \boxed{}°$

d *d*

$d = \boxed{}°$

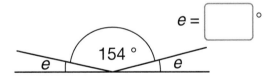
154°
e *e*

$e = \boxed{}°$

5 Half a cake is divided into three slices that are different sizes.

Slice 2 is three times of size of Slice 1. Slice 3 is twice the size of Slice 1.

Use the diagram to work out the size of the angles made by each of the three slices.

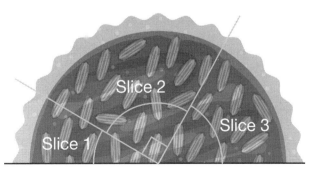
Slice 2
Slice 3
Slice 1

Slice 1 = $\boxed{}$ Slice 2 = $\boxed{}$ Slice 3 = $\boxed{}$

Date: _____

Lesson 1: **Perimeter of simple 2D shapes**

- Measure and calculate the perimeter of simple 2D shapes

You will need
- paper for working out
- ruler

1 Find the perimeter of each shape. Shape **b** is a square.

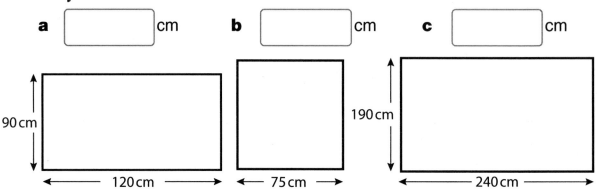

a [ ] cm   **b** [ ] cm   **c** [ ] cm

90 cm   120 cm
75 cm
190 cm   240 cm

2 Calculate the missing dimension of each shape given its perimeter and one dimension. Remember to estimate first, e.g. part **d**. Double the missing dimension is given by 76 subtract double 26. As 26 is close to 25, 76 subtract double 25 (50) is 26. The answer will be close to 13 (half 26).

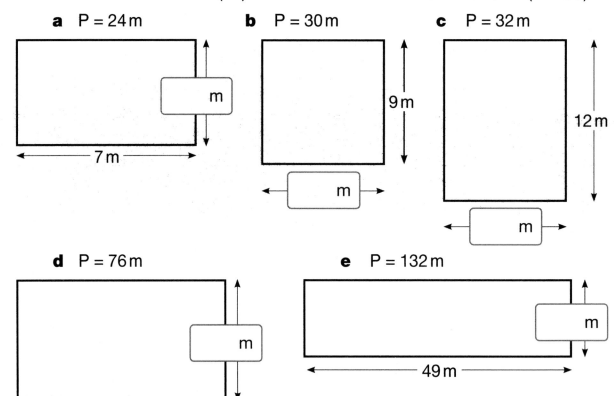

a P = 24 m   **b** P = 30 m   **c** P = 32 m

m   9 m   12 m
7 m   m   m

d P = 76 m   **e** P = 132 m

m   m
26 m   49 m

3 Draw and label rectangles A, B and C on the grid so they have the given perimeters. Assume that each square on the grid is one square centimetre.

Rectangle A:
P = 28 cm
l = 8 cm

Rectangle B:
P = 44 cm
l = 17 cm

Rectangle C:
P = 52 cm
l = 18 cm

Remember!
l stands for length.

Geometry and Measure

4 Calculate the answers to these problems.

a The length (l) of a rectangular field is double its width (w). It has a perimeter of 54 m. What are the dimensions of the field?

l =

w =

b The length of a rectangular playground is three times its width. It has a perimeter of 88 m. What are the dimensions of the playground?

l =

w =

c The length of a rectangular window is five times its width. It has a perimeter of 396 cm. What are the dimensions of the window?

l =

w =

Date: _____

Lesson 2: **Perimeter of compound 2D shapes**

* Calculate the perimeter of compound 2D shapes

You will need
* paper for working out
* ruler

1 Calculate the perimeter of each shape by finding the sum of all the sides. The shapes are not drawn to scale.

a [ ] cm   **b** [ ] cm   **c** [ ] cm

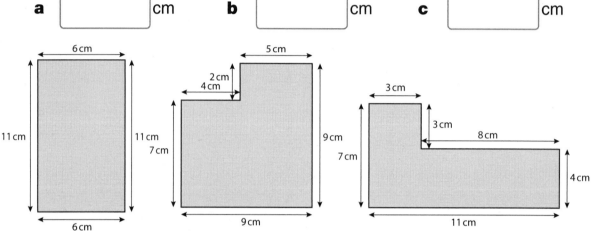

2 Work out the perimeter of each shape. Use any method you prefer. The shapes are not drawn to scale. Remember to estimate first. e.g. Part a. Since both 11m and 9m are close to 10, the perimeter will be close to 2 × 10 add 2 × 10 = 40.

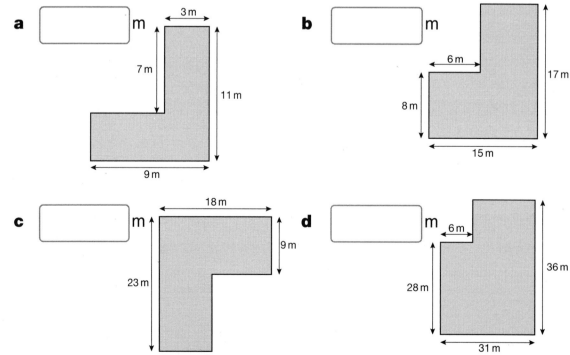

a [ ] m

b [ ] m

c [ ] m

d [ ] m

Geometry and Measure

3 Find the perimeter of each shape by thinking of it as a single rectangle. The larger rectangle has been completed for shape **a**.

a [ ] m

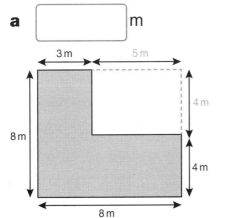

3 m   5 m
8 m   4 m
4 m
8 m

b [ ] m

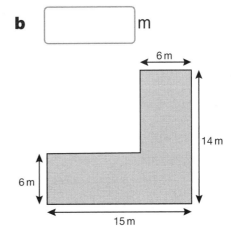

6 m
14 m
6 m
15 m

c [ ] m

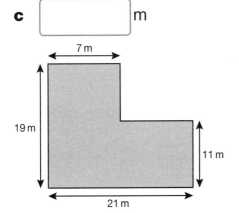

7 m
19 m
11 m
21 m

d [ ] m

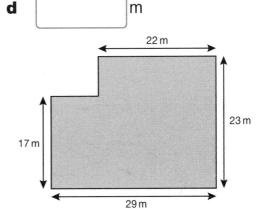

22 m
17 m   23 m
29 m

4 Calculate the perimeter of each shape by thinking of each shape as a complete rectangle.

a [ ] cm

b [ ] cm

c [ ] cm

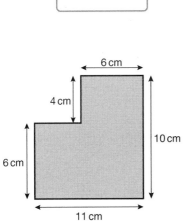

6 cm
4 cm
10 cm
6 cm
11 cm

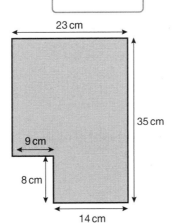

23 cm
35 cm
9 cm
8 cm
14 cm

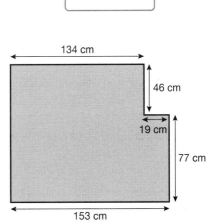

134 cm
46 cm
19 cm
77 cm
153 cm

Date: _____

Lesson 3: **Area of simple 2D shapes**

Geometry and Measure

- Measure and calculate the area of simple 2D shapes
- Understand that shapes with the same perimeter can have different areas and vice versa

You will need

- paper for working out
- ruler
- squared paper

1 Use the rule to calculate the area of each shape.
Give your answer in square centimetres. The shapes are not drawn to scale. Remember to estimate first, e.g. part **d**. 25 is close to 30 so the answer will be close to $80 \times 30 = 2400$.

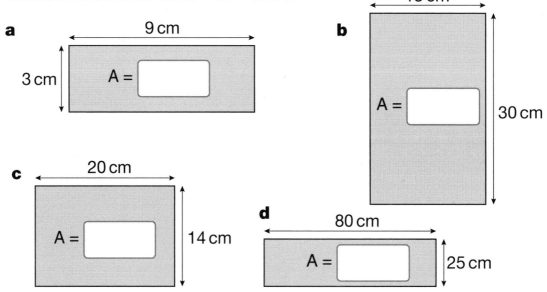

a 9 cm   3 cm   A =

b 15 cm   A =   30 cm

c 20 cm   A =   14 cm

d 80 cm   A =   25 cm

2 You are given the area and one dimension of each shape. Fill in the missing dimension. The shapes are not drawn to scale.

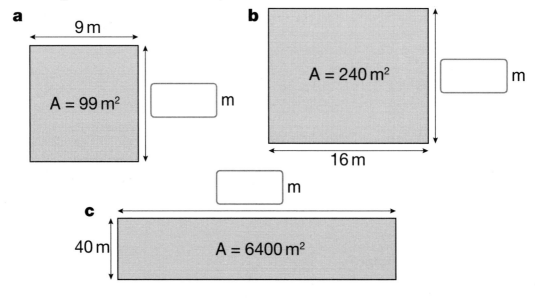

a 9 m $A = 99\,m^2$ ___ m

b $A = 240\,m^2$ ___ m   16 m

c ___ m   40 m $A = 6400\,m^2$

3 Calculate the answers to these problems.

a The length of a rectangular lawn is three times its width. The perimeter of the rectangle is 72 m. What is the area of the lawn?

b A rectangular piece of cardboard is 5 times as long as it is wide. The area is 180 square metres. What is its perimeter?

4 Three children are given the same shape to draw.

Alpesh says:

Lakshmi says:

Sunita says:

The smallest length is 2 cm.

The perimeter is 24 cm.

The area is less than 40 cm².

Use squared paper to show the shapes that each of the children might have drawn.

Date: _____

Lesson 4: **Area of compound 2D shapes**

• Calculate the area of compound 2D shapes

You will need

• paper for working out
• ruler

1 Each square on this grid represents 1 cm². Find the area of each shaded shape. Give your answer in square centimetres.

A:

B:

C:

D:

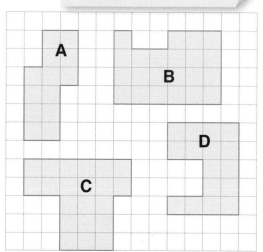

2 Calculate the area of these shapes by dividing them into individual rectangles. The first two shapes have been divided to help you.

a [ ] m²

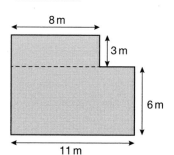

8 m
3 m
6 m
11 m

b [ ] m²

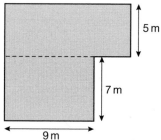

14 m
5 m
7 m
9 m

c [ ] m²

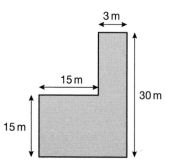

3 m
15 m
30 m
15 m

d [ ] m²

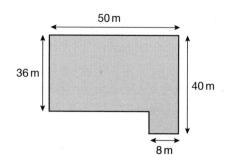

50 m
36 m
40 m
8 m

Geometry and Measure

3 Use the subtraction method to calculate the area of each shape.
The first one has been done for you.

A $= (15\,m \times 10\,m) - (3\,m \times 7\,m)$

$= 150\,m^2 - 21\,m^2$

$= 129\,m^2$

B _____

_____

_____

_____

C _____

_____

_____

_____

D _____

_____

_____

_____

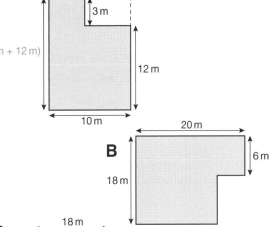

A
3 m   7 m (10 m − 3 m)
3 m
15 m (3 m + 12 m)
12 m
10 m

B
20 m
6 m
18 m
15 m

C
18 m
17 m   21 m
25 m

D   7 m
35 m
18 m
16 m

4 Work out the area of the grey shading in the diagram.

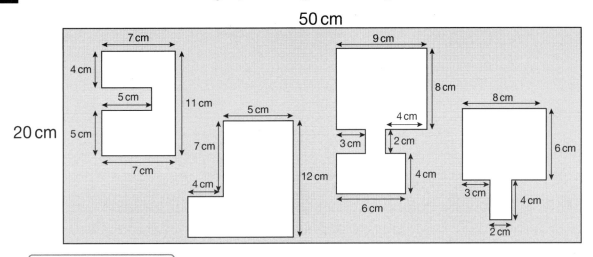

50 cm

7 cm
4 cm
5 cm
11 cm
5 cm
7 cm
7 cm
20 cm
5 cm
5 cm
12 cm
4 cm
9 cm
8 cm
4 cm
3 cm   2 cm
6 cm   4 cm
8 cm
6 cm
3 cm   4 cm
2 cm

Date: _____

189

Lesson 1: **Comparing coordinates (1)**

• Compare two points plotted on the coordinate grid to say which is closer to each axis

1 Plot each point and write its coordinates.

a A point **C** that is closer to the x-axis than point **D**.

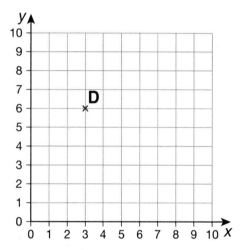

Point C: (__, __)

b A point **E** that is further away from the x-axis than point **F**.

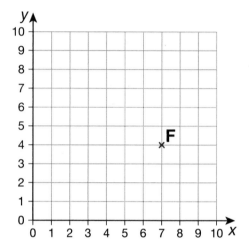

Point E: (__, __)

c A point **H** that is closer to the y-axis than point **G**.

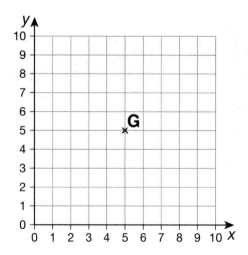

Point H: (__, __)

d A point **K** that is further away from the y-axis than point **J**.

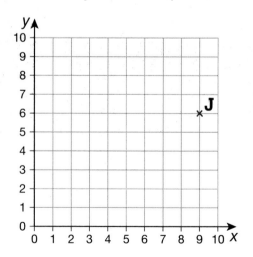

Point K: (__, __)

Geometry and Measure

2 Plot the points on the coordinate grid. Then write the letters in the correct column in the table.

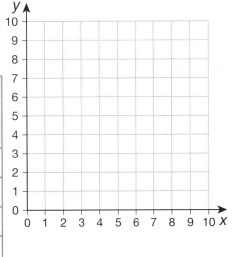

Point	Point	Point further away from the *x*-axis	Point closer to the *y*-axis
A (3, 7)	B (2, 9)		
C (5, 6)	D (1, 2)		
E (7, 8)	F (9, 6)		
G (4, 10)	H (6, 6)		
I (9, 4)	J (10, 7)		

3 Complete the table with coordinates for the points described.

Point	A point closer to the *x*-axis	A point further away from the *y*-axis
(4, 7)		
(2, 6)		
(9, 5)		
(2, 8)		
(6, 3)		

4 Complete the table. You must do this without plotting points on a coordinate grid.

Point	Point	A point further away from the *x*-axis	A point closer to the *y*-axis
A (4, 8)	B (3, 7)		
C (8, 5)	D (9, 4)		
E (16, 3)	F (17, 2)		
G (19, 12)	H (15, 8)		
I (24, 16)	J (29, 18)		

Date: _____

Geometry and Measure

Lesson 2: **Comparing coordinates (2)**

• Without a grid, estimate the position of point A relative to point B given the coordinates of point B

You will need
• ruler

1 Plot each point and write its coordinates.

a A point **C** that is closer to the *x*-axis than point **D**.

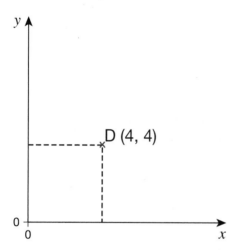

Point C: (__, __)

b A point **E** that is further away from the *x*-axis than point **F**.

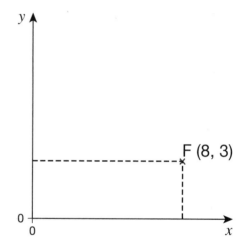

Point E: (__, __)

c A point **H** that is closer to the *y*-axis than point **G**.

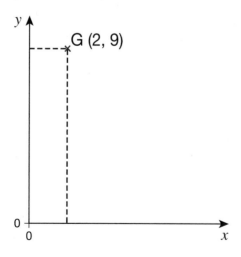

Point H: (__, __)

d A point **K** that is further away from the *y*-axis than point **J**.

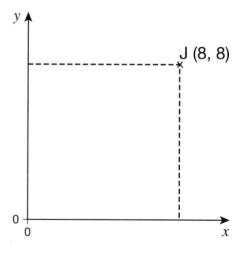

Point K: (__, __)

Geometry and Measure

2 Plot each point.

a Point C (8, 2)

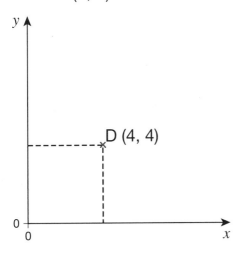

b Point E (4, 6)

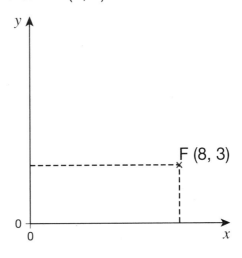

c Point H (1, 5)

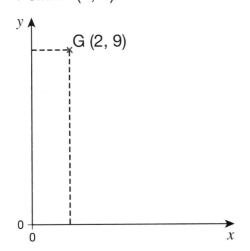

d Point K (5, 5)

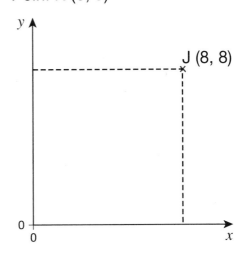

3 Compare the coordinates of each pair of points. Write the difference between the values.

Point	Point	Difference between the x-coordinate values	Difference between the y-coordinate values
A (2, 7)	B (4, 8)		
C (13, 8)	D (10, 5)		
E (9, 9)	F (16, 3)		
G (20, 11)	H (14, 7)		

Date: _____

193

Lesson 3: **Plotting coordinates (1)**

Geometry and Measure

- Plot points to form squares in the first quadrant

You will need
- ruler

1 Use the coordinates to plot each square.

a (2, 2) (2, 4)
(4, 4) (4, 2)

b (1, 3) (6, 3)
(6, 8) (1, 8)

c (8, 0) (8, 8)
(0, 8) (0, 0)

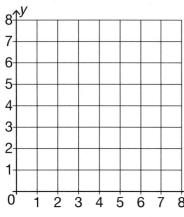
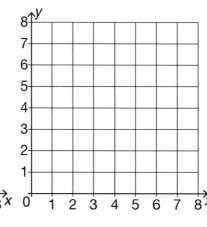
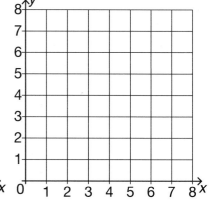

2 Draw each square described. Write the coordinates of the vertices.

a Start at (2, 3).
Length of sides:
4 units

b Start at (7, 5).
Length of sides:
5 units

c Start at (1, 8).
Length of sides:
7 units

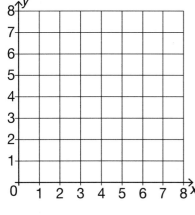
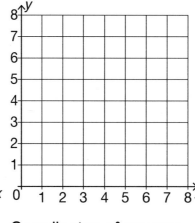
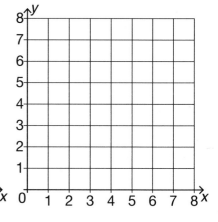

Coordinates of square:

(2, 3) (__, __)
(__, __) (__, __)

Coordinates of square:

(7, 5) (__, __)
(__, __) (__, __)

Coordinates of square:

(1, 8) (__, __)
(__, __) (__, __)

Geometry and Measure

3 Draw the square and identify the coordinates of the missing vertex.

a Square ABCD
A (2, 2), B (2, 7), C (7, 7)

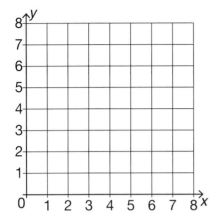

Coordinates of D: (__, __)

b Square EFGH
E (8, 1), F (8, 8), G (1, 8)

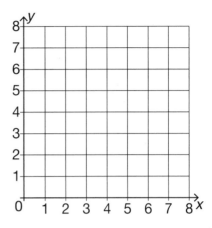

Coordinates of H: (__, __)

4 Calculate the coordinates of the missing vertex of each square.

Coordinates of three vertices of a square	Coordinates of fourth vertex
A (4, 4), B (7, 4), C (7, 7)	D (__, __)
E (3, 8), F (8, 8), G (3, 3)	H (__, __)
J (9, 10), K (2, 10), L (2, 3)	M (__, __)
N (4, 0), O (9, 0), P (4, 5)	Q (__, __)
R (8, 1), S (2, 7), T (8, 7)	U (__, __)

5 Points A and B mark two vertices of a square ABCD.

Plot possible coordinates for vertices C and D on the diagram. Write the coordinates.

C: (__, __)   D: (__, __)

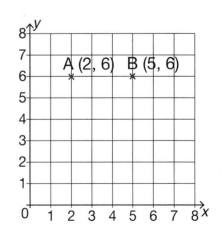

Date: _____

Lesson 4: **Plotting coordinates (2)**

• Plot points to form shapes in the first quadrant

You will need
• ruler

1 Use the coordinates to plot each shape.

a (3, 3), (8, 3),
(8, 8), (3, 8)

b (2, 6), (7, 6),
(7, 3)

c (1, 1), (1, 6),
(8, 6), (8, 1)

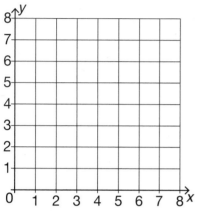
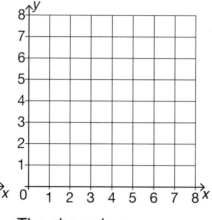
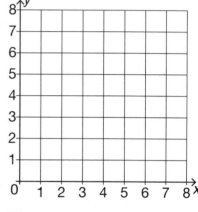

The shape is a

_____.

The shape is a

_____.

The shape is a

_____.

2 Draw each shape on the grid. Write the coordinates of each vertex of the shape.

a Draw a right-angled triangle.
Start at point (5, 2).

b Draw a square.
Start at point (7, 7).

c Draw a rectangle.
Start at point (0, 6).

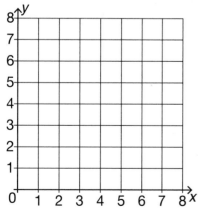
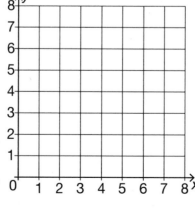
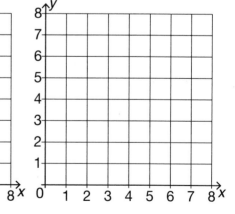

Coordinates of triangle:

(5, 2) (__, __) (__, __)

Coordinates of square:

(7, 7) (__, __) (__, __)
(__, __)

Coordinates of rectangle:

(0, 6) (__, __) (__, __)
(__, __)

3 Draw each shape and identify the coordinates of the missing vertex.

a Plot these points:
A (2, 2) B (2, 8)
C (7, 8).
ABCD is a
rectangle.

b Plot these points:
A (2, 7) B (8, 5).
ABC is an
isosceles triangle.

c Plot these points:
A (1, 4) B (4, 7)
C (7, 4)
ABCD is a
square.

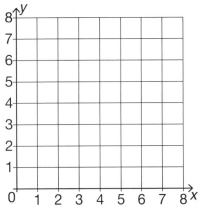
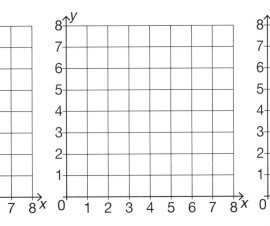
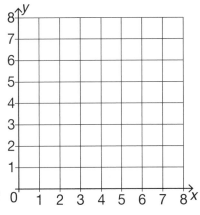

What are the coordinates of D? (__, __)

What are the coordinates of C? (__, __)

What are the coordinates of D? (__, __)

4 Draw each shape and identify the coordinates of the missing vertex.

a Plot these points: A (4, 3)
B (1, 2) C (1, 6) D (4,7) E (7, 6)
ABCDEF is a
symmetrical hexagon.

b Plot these points:
A (2, 8) B (6, 7) C (7, 5) D (6, 3)
ABCDE is a
symmetrical pentagon.

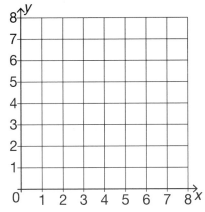
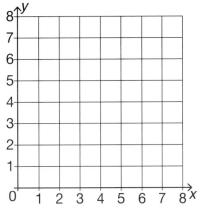

What are the coordinates of F? (__, __)

What are the coordinates of E? (__, __)

Date: _____

197

Geometry and Measure

Lesson 1: **Describing translations**

Geometry and Measure

- Describe the translation of a 2D shape on a square grid

You will need
- ruler

1 Cross through the movements that are **not** translations.

a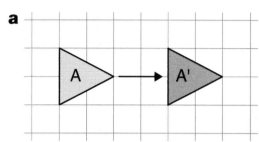

b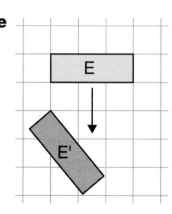

c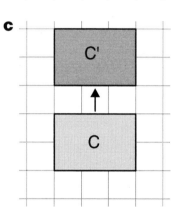

d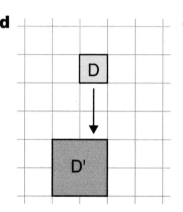

e

2 Complete each sentence.

a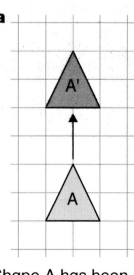

Shape A has been

translated ☐

squares ☐.

b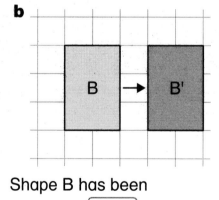

Shape B has been

translated ☐ squares

☐.

c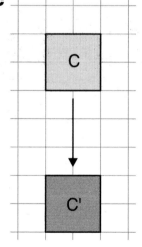

Shape C has been

translated ☐

squares ☐.

Geometry and Measure

3 Describe the translation of each shape from A to B.

a

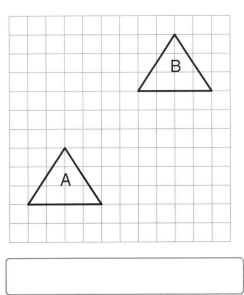

b

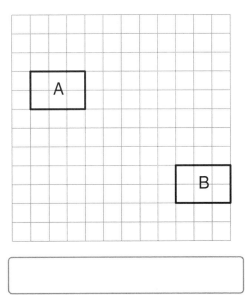

4 Shape A has been translated 4 times. Shapes B, C, D and E show how Shape A has been translated. Describe each translation next to it.

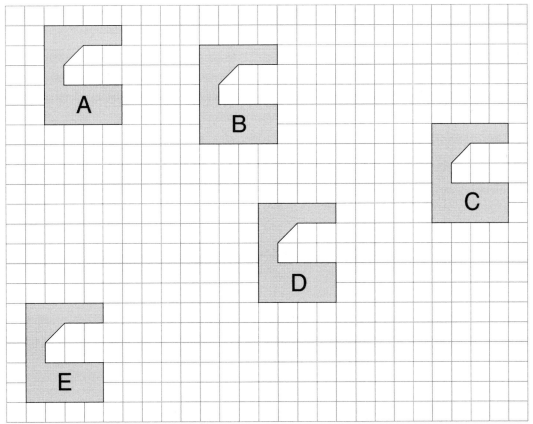

Date: _____

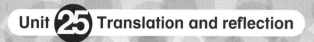
Lesson 2: **Performing translations**

Geometry and Measure

- Perform the translation of a 2D shape on a square grid

You will need
- ruler
- coloured pencils

1 Translate each shape as described.

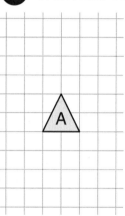 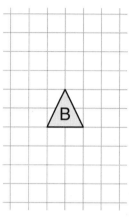 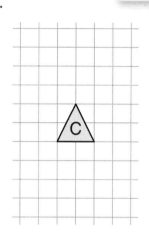 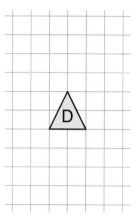

| a | Left 2 squares | b | Down 4 squares | c | Right 1 square | d | Up 4 squares |

2 Translate each shape as described.

a

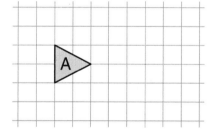

Right 5 squares, up 1 square

b

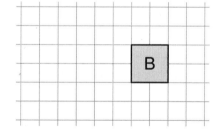

Down 1 square, left 4 squares

c

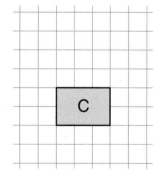

Up 3 squares, right 1 square

d

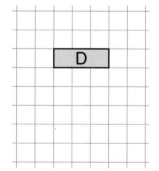

Down 4 squares, left 1 square

3 Translate each shape.

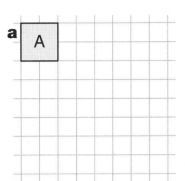

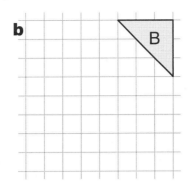

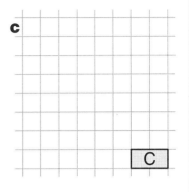

Translate:
- right 5
- down 4
- left 3

Translate:
- left 5
- down 5
- right 4

Translate:
- up 5
- left 6
- down 4

4 Make patterns by translating each shape.

a

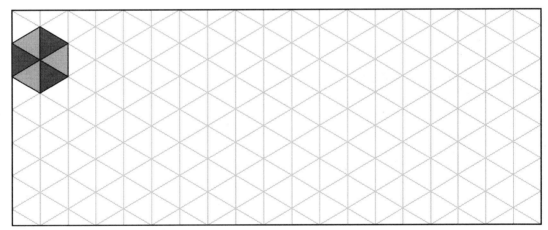

b

Date: _____

201

Lesson 3: **Reflecting shapes (1)**

Geometry and Measure

- Reflect 2D shapes in both horizontal and vertical mirror lines on square grids

You will need
- squared paper
- mirror
- coloured pencils

1 Place a mirror along the line of symmetry. Draw the reflection of the shape.

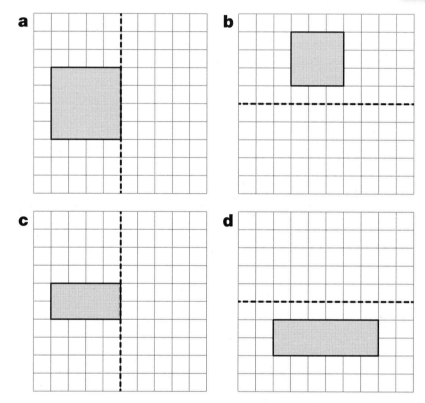

2 Reflect each shape in the mirror line.

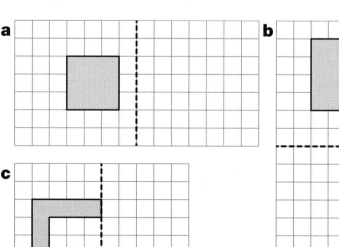

Geometry and Measure

3 Reflect each of the irregular shapes in the mirror line.

a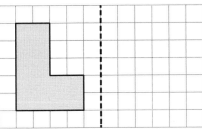

b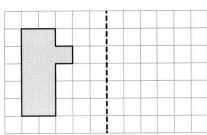

c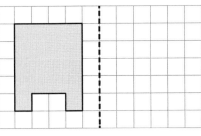

d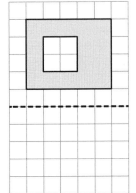

e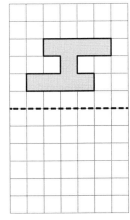

f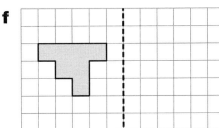

4 Simple shapes can be made into a design using reflections.

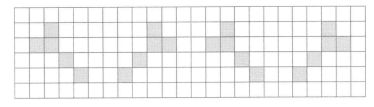

Using square paper, make a simple shape by colouring several squares in the same colour. Then reflect your shape several times to create a design.

Write about how you created your design.

_____

_____

_____

Date: _____

Lesson 4: **Reflecting shapes (2)**

- Predict and draw where a shape will be after reflection where the sides of the shape are not vertical or horizontal

You will need
- ruler

1 Reflect the line in the mirror line.

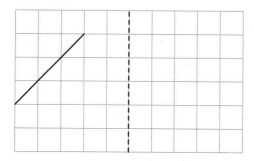

2 Mark in the vertices of each shape and reflect them in the mirror line. Connect up the dots to form the reflected image.

a

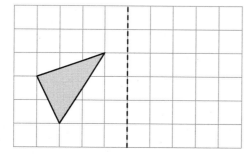

b

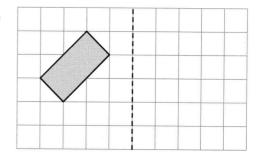

c

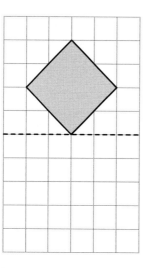

d

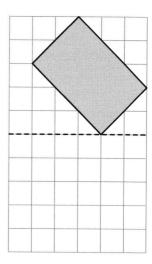

3 Reflect each irregular polygon and draw the image.

a
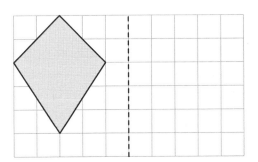

b
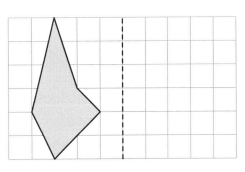

c
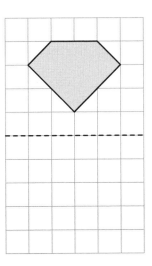

d
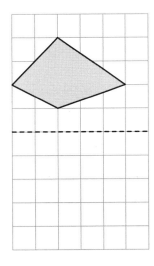

4 Reflect the polygon and draw the image.

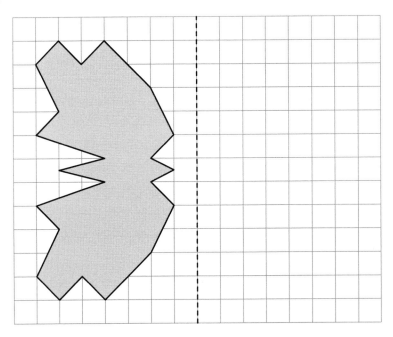

Date: _____

Geometry and Measure

Lesson 1: **Venn and Carroll diagrams**

Statistics and Probability

- Know how to construct a statistical question
- Represent data in Venn and Carroll diagrams

You will need
- paper

1 Drake Class investigated the question: *How many learners in our class have sisters only?*

How many learners in Drake Class have:

a sisters only? ☐

b brothers only? ☐

c both brothers and sisters? ☐

d no brothers or sisters? ☐

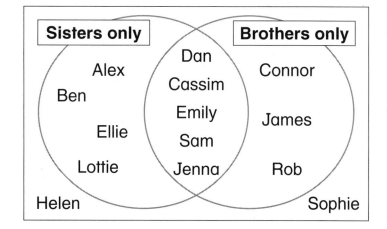

2 The learners in Darwin Class collected data to investigate the question: *How many learners in our class are taller than 125 cm?* The table below shows the data collected for ten learners.

Name (boy, girl)	Will (B)	Leila (G)	Blake (B)	Kamari (G)	Marco (B)	Libby (G)	Ashton (B)	Mia (G)	Ibrahim (B)	Aisha (G)
Height (cm)	127	122	130	132	124	121	131	126	122	129

Write the names in the correct part of the Carroll diagram. Use the information to answer the questions that follow.

	Boy	**Not boy**
125 cm or shorter		
Not 125 cm or shorter		

How many of the learners are:

a girls taller than 125 cm? ☐

b boys 125 cm or shorter? ☐

c boys taller than 125 cm? ☐

d girls 125 cm or shorter? ☐

e 125 cm or shorter? ☐

f taller than 125 cm? ☐

3 **a** Write two statistical questions that would provide data that could be represented by a Venn diagram.

i _____

ii _____

b Write two statistical questions that would provide data that could be represented by a Carroll diagram.

i _____

ii _____

4 **a** Write a statistical question that can be answered by collecting data from learners in your class. The data should be represented in a Venn or Carroll diagram.

_____

b Describe how you would collect the data.

_____

c Use a separate piece of paper to collect the data. Represent the data in a Venn or Carroll diagram.

d Answer the question posed in **4 a** by interpreting the data you have collected.

_____

_____

_____

_____

Date: _____

Lesson 2: **Tally charts, frequency tables and bar charts**

Statistics and Probability

* Know how to construct a statistical question
* Represent data in tally charts, frequency tables and bar charts

You will need
* coloured pencil

1 Year 5 voted for their favourite non-fiction book subject. The tally chart shows the results. Complete the frequency column. Then complete the bar chart using thedata in the table.

Subject	Votes	Frequency
history	卌 /	
space	卌 卌 //	
sport	////	
art	卌 卌 ////	
nature	卌 卌 卌 /	

Favourite non-fiction books

Number of children (y-axis: 0 to 18)

Subject (x-axis: history, space, sport, art, nature)

a Which subject was the most popular?

_____

b Which subject was the least popular? _____

c How many more learners voted for nature than history? ☐

d How many learners voted in the survey? ☐

2 20 learners recorded their scores in a game.

Score	Tally	Frequency (*f*)	Percentage
1	////	4	20%
2	卌 /	6	30%
3	卌 //	7	35%
4	/	1	5%
5	//	2	10%

a What percentage of learners scored 1 or 2 points? ☐

b What percentage of learners scored 3 or more points? ☐

Statistics and Probability

3 These are the masses, in kilograms, of 50 children. Complete the grouped tally chart for the given data. Data values that are multiples of 10 should be recorded in the lower group range.

23	68	17	45	35	21	27	34	59	60
42	49	34	56	15	16	32	50	28	37
54	57	42	57	64	56	68	37	66	25
33	26	25	23	63	43	19	36	55	37
28	61	44	58	52	47	69	52	41	39

Mass (kg)	Tally	Frequency
10–20		
20–30		
30–40		
40–50		
50–60		
60–70		

a How many children have a mass between 30 kg and 40 kg? ☐

b How many children have a mass of 50 kg or less? ☐

c Write two other conclusions you can draw from the data.

_____

_____

4 These are the numbers of birds in a garden. They are recorded at the same time each day over a 50-day period.

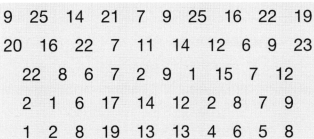

9	25	14	21	7	9	25	16	22	19
20	16	22	7	11	14	12	6	9	23
22	8	6	7	2	9	1	15	7	12
2	1	6	17	14	12	2	8	7	9
1	2	8	19	13	13	4	6	5	8

a On a separate piece of paper, construct a tally chart similar to the one in **3** , using groups of 1–5, 6–10, 11–15, 16–20, 21–25. Include and complete a column for 'frequency'.

b What conclusions can you draw from the data? _____

_____

Date: _____

Lesson 3: **Waffle diagrams**

- Know how to construct a statistical question
- Represent data in waffle diagrams

You will need
- two waffle diagrams
- coloured pencils

1 100 people were asked to name their favourite sport. Use the waffle diagram to complete the frequency table.

Favourite sport	Frequency
football	
rugby	
tennis	35
netball	
hockey	

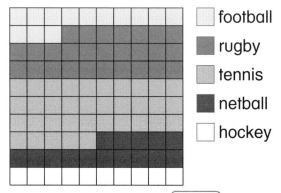

football
rugby
tennis
netball
hockey

a How many more people voted for tennis than rugby? ☐

b How many fewer people voted for hockey than netball? ☐

2 100 learners in a school were asked the question, 'What is your favourite sandwich filling?' The data was recorded in a frequency table. Represent the data in a waffle diagram on a piece of paper. The waffle diagram will need a title and a colour-coded key.

Filling	Frequency	Percentage
cheese	37	37%
egg	19	19%
tuna	42	42%
hummus	2	2%

What conclusions can you draw from the data?

_____

_____

_____

_____

Statistics and Probability

3 These are the numbers of hours Tom spent reading every week for 50 weeks.

21	18	4	31	6	19	23	7	15	24
33	26	2	9	14	17	11	8	20	27
24	32	6	10	14	28	20	32	2	19
29	16	15	17	25	12	14	22	23	4
5	7	19	34	27	22	7	11	24	18

a Complete the frequency table. Assume that any data that is a multiple of 10 will be entered in the lower interval group.

Number of hours	Tally	Frequency
1–10		
10–20		
20–30		
30–40		

b Use the values from your completed frequency table to draw a waffle diagram. Remember to give it a title and a key.

c i How many weeks did Tom read for 20 or more hours?

ii How many weeks did Tom read for 30 or hours or less?

d What conclusions can you draw from the data?

_____

_____

_____

_____

_____

_____

Date: _____

Lesson 4: **Mode and median**

- Find and interpret the mode and the median of a data set

1 Tom did a survey of his friends' favourite ice cream flavours. He put the results in a table.

Friend	Favourite flavours
Amir	chocolate, vanilla, bubblegum
Florence	strawberry, vanilla, cookie dough
Jack	bubblegum, cookie dough, chocolate
Ashia	cookie dough, strawberry
Ethan	vanilla, chocolate, cookie dough

Count how many there are of each flavour.

chocolate	
vanilla	
bubblegum	
cookie dough	
strawberry	

a The flavour that appears most is _____.

b The mode is _____ (ice cream flavour).

2 Find the mode of each data set.

a 3, 13, 16, 3, 12, 8, 22, 16, 8, 6, 3   Mode:

b 17, 7, 3, 12, 3, 5, 5, 12, 10, 24, 12   Mode:

c apple, banana, apple, pear, peach, banana, peach, apple   Mode:

d red, blue, green, pink, pink, green, red, blue, pink, green, pink, red, pink   Mode:

e B, X, P, R, C, C, R, P, R, B, X, R, C, X, R   Mode:

f 13 cm, 5 cm, 9 cm, 3 cm, 11 cm, 13 cm, 9 cm, 3 cm, 9 cm, 9 cm, 5 cm, 13 cm   Mode:

3 Find the median of each set of scores.

a

Football scores
4, 3, 1, 2, 5

Median [ ]

b

Hockey scores
6, 4, 8, 9, 5, 6, 4

Median [ ]

c

Baseball scores
4, 7, 6, 8, 3, 5, 7, 8, 6

Median [ ]

d

Video game scores
77, 66, 49, 58, 75

Median [ ]

4 The table shows the average high and low temperatures recorded by a class over a period of 12 months.

Month	High	Low
January	10°	1°
February	12°	2°
March	16°	3°
April	22°	5°
May	26°	10°
June	30°	14°
July	16°	15°
August	31°	16°
September	28°	10°
October	23°	6°
November	16°	2°
December	11°	3°

a What is the mode of each set of temperatures?

High [ ]   Low [ ]

b How would the mode change if the first four low temperatures from January to April were:

1°, 2°, 3° and 2°? [ ]

c In which months was the average high temperature below 15°?

_____

d What is the median temperature of the months with an average high temperature above 15°? [ ]

Statistics and Probability

Lesson 1: **Frequency diagrams and line graphs**

- Know how to construct a statistical question
- Represent data in frequency diagrams and line graphs

You will need
- squared paper
- ruler
- coloured pencil

 The average air temperature was recorded throughout the year. A line graph was drawn to display the data. Use the information to answer the questions.

a What was the temperature in February? ☐

b In which two months was the temperature 20 °C?

☐ and ☐

c How much higher was the temperature in June than October? ☐

d How much lower was the temperature in March than July? ☐

e Between which two months did the temperature rise the fastest?

☐ and ☐

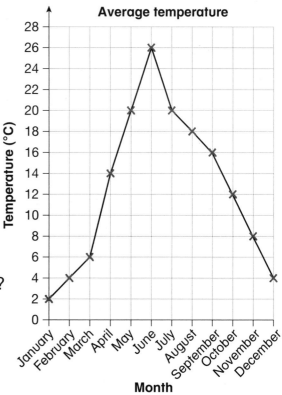

Average temperature

2 The data lists the ages of people attending a gym.

54	33	34	23	41	27	48	31	32
18	51	61	41	17	35	22	44	46
39	44	36	15	32	67	27	29	62
27	22	37	58	21	56	26	38	32

a Record the data in the frequency table. Then draw a line graph on squared paper to display the data.

Age (years)	Frequency
10–20	
20–30	
30–40	
40–50	
50–60	
60–70	

b Write three facts you can read from your graph.

i _____

ii _____

iii _____

3 Jessica played a board game 30 times and recorded her scores. She then drew a frequency diagram to display the results.

Frequency of board game scores

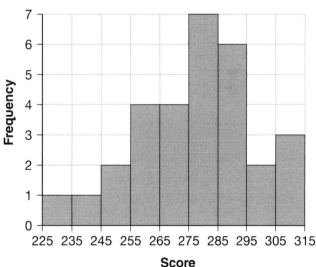

a How many games did she score 275 or under?

b How many games did she score 275 or over?

c Which score interval contains the median score?

Explain how you worked this out.

_____

_____

Date: _____

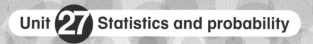

Lesson 2: **Dot plots**

- Know how to construct a statistical question
- Represent data in dot plots

You will need
- squared paper
- ruler
- coloured pencil

1 A count was made of the number of ladybirds per leaf for 30 leaves. The data was used to draw a dot plot.

  a How many leaves had 8 ladybirds?

  b Which number of ladybirds was only found on 3 leaves?

  c How many leaves had over 9 ladybirds?

  d How many leaves had fewer than 8 ladybirds?

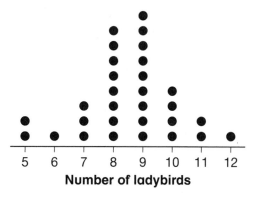

Frequency of ladybirds on a leaf

Number of ladybirds

2 A supermarket manager opened several boxes of chocolates of the same type. He wanted to see if the number of chocolates was the same in each box. He counted and recorded the numbers and then constructed a dot plot to represent the data. Use the graph to answer the questions.

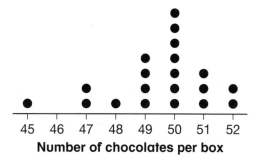

Number of chocolates

Number of chocolates per box

  a Which number of chocolates was most frequent?

  b How many boxes contained from 49 to 51 chocolates? (Include 49 and 51.)

  c How many boxes were opened altogether?

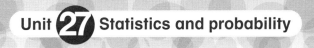

Statistics and Probability

3 Jo picked a random page in a book 36 times. For each page, he counted and recorded the number of lines of text.

32	39	36	35	30	34	31	29	34
36	38	32	34	36	40	35	32	31
35	41	42	31	30	39	34	38	40
38	29	36	34	35	32	31	30	34

a Draw a tally chart or frequency table for this data.

b Use squared paper to draw a dot plot to display the data.

c Write three facts you can read from your dot plot.

   i _____

   ii _____

   iii _____

4 Wendall Class counted the number of petals on flowers in one area of a meadow. A dot plot was constructed from the data collected.

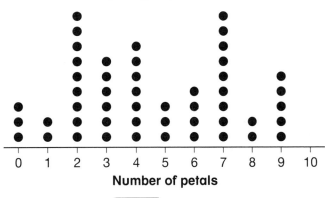

Frequency of petals found

Number of petals

a How many flowers had 2 or fewer petals?

   [ ]

b How many flowers had 7 or more petals? [ ]

c What are the modal frequencies? [ ]

d Which is the median number of petals? [ ]

Explain how you worked this out.

_____

_____

_____

Date: _____

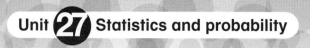

Lesson 3: **Probability (1)**

- Describe the likelihood of an event happening, using the language of chance

1 For each event, tick the likelihood of it happening.

Event	Likelihood		
	impossible	possible	certain
pick a blue counter from a bag of red counters			
pick a counter from a bag of counters			
pick a yellow counter from a mixed bag of green and yellow counters			
pick a green counter from a bag of green counters			

2 Label the spinners with numbers 1, 2, 3 or 4 to satisfy the probability rules given.

a

b

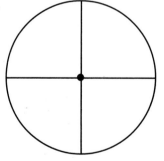

c

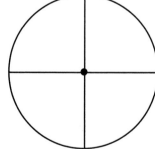

It is **possible** to spin 2, 3 or 4.

It is **impossible** to spin 1.

Spinning 4 is **certain**.

There is an **even chance** of spinning 2 or 3.

Statistics and Probability

Statistics and Probability

3 Look at the letters on the spinner. Tick the boxes that apply.

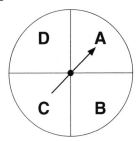

Event	Probability		
	impossible	**possible**	**certain**
spinning **B**			
spinning **C**			
spinning **E**			
spinning a letter			

4 Draw a line from each event to where you think it should be classified on the scale.

You toss a coin and it lands on heads.	You roll a 1–6 dice and get a 6.	You flap your arms and fly up in the sky.	Your age will increase each year.	The next vehicle you see will have 4 wheels.

```
├──────────────┼──────────────┼──────────────┼──────────────┤
impossible    unlikely      even chance      likely      certain
```

5 A jar contains 4 red counters, 7 green counters, 3 blue counters and 4 yellow counters. A counter is taken at random from the jar. Compare the probabilities by writing >, < or = between the statements.

a probability of taking a green counter ☐ probability of taking a red counter

b probability of taking a blue counter ☐ probability of taking a red counter

c probability of taking a green counter ☐ probability of taking a blue counter

d probability of taking a red counter ☐ probability of taking a yellow counter

Date: _____

Lesson 4: **Probability (2)**

• Describe the results of chance experiments, using the language of probability

You will need
• coloured pencils
• blank cards

1 A coin is flipped.

5

a Describe the chance of flipping 'tails'.

b Describe the chance of flipping 'heads'.

c Describe the chance of flipping a different design.

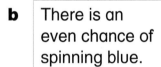

2 Colour each spinner using the colours red, green, blue and yellow only to give the chance or chances described.

a | It is possible to spin red, blue or yellow.

It is impossible to spin green.

b | There is an even chance of spinning blue.

There is an even chance of spinning yellow.

c | The chance of spinning red is likely.

The chance of spinning blue is unlikely.

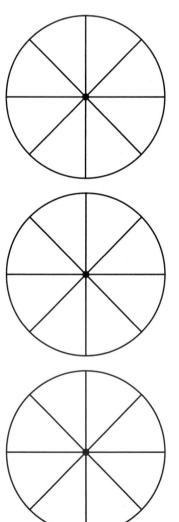

Statistics and Probability

 You are given a pack of 8 digit cards. There is more than one of each card. Design each pack to satisfy the rule given. Write the digit in the corner of the card.

Pack 1: It is certain to draw a '4'.

Pack 2: There is an even chance of drawing a '3'.

Pack 3: It is likely to draw a '9' and unlikely to draw a '2'.

Pack 4: It is likely to draw a '2', unlikely to draw a '6' and possible to draw a '3'.

Conduct experiments with the four packs you have designed. For each pack, draw 20 times and record the result. Remember to replace the card and shuffle the pack each time. Does the frequency of the digits drawn agree with the probabilities predicted for each pack? How do you know?

Pack 1: _____　　　Pack 2: _____

Pack 3: _____　　　Pack 4: _____

If they are different, how would you explain this? _____

4 Amani places the beads on the right in a bag. A bead
5 is drawn from the bag at random. Describe the probability of drawing a bead from the bag in as many ways as you can, including fractions and ratios.

a A cube _____

b A sphere _____

c A cylinder _____

Date: _____

Acknowledgements

Photo acknowledgements

Every effort has been made to trace copyright holders. Any omission will be rectified at the first opportunity.

p12l Gomolach/Shutterstock; p12r SingjaiStock/Shutterstock; p17 Afaf.asf/Shutterstock; p20a Lomonovskyi/Shutterstock; p20b Mega Pixel/Shutterstock; p20c Alexlukin/Shutterstock; p20d Voyata/Shutterstock; p20e E_Vector/Shutterstock; p20f Delices/Shutterstock; p20g Pretty Vectors/Shutterstock; p20h Tkray/Shutterstock; p20i Pretty Vectors/Shutterstock; p20j FocusStocker/Shutterstock; p20k Michael Kraus/Shutterstock; p20l Kostenko Maxim/Shutterstock; p20m M. Unal Ozmen/Shutterstock; p20n Pixfiction/Shutterstock; p20o AVIcon/Shutterstock; p21tl FocusStocker/Shutterstock; p21tc Pixfiction/Shutterstock; p21tr Michael Kraus/Shutterstock; p21bl AVIcon/Shutterstock; p21bc Kostenko Maxim/Shutterstock; p21br M. Unal Ozmen/Shutterstock; p28a Lomonovskyi/Shutterstock; p28b Alexlukin/Shutterstock; p28c E_Vector/Shutterstock; p28d Mega Pixel/Shutterstock; p28e Voyata/Shutterstock; p28f Delices/Shutterstock; p28g Tkray/Shutterstock; p28h Pretty Vectors/Shutterstock; p28i Pretty Vectors/Shutterstock; p28j Sudowoodo/Shutterstock; p28k Pambudi/Shutterstock; p28l Farah Sadikhova/Shutterstock; p28m VectorPlotnikoff/Shutterstock; p28n Edel/Shutterstock; p29l Pambudi/Shutterstock; p29cl Edel/Shutterstock; p29cr Farah Sadikhova/Shutterstock; p29r VectorPlotnikoff/Shutterstock; p53 Titov Nikolai/Shutterstock; p60l Webicon/Shutterstock; p60t Olegtoka/Shutterstock; p60b Vahe 3D/Shutterstock; p61 Titov Nikolai/Shutterstock; p104 ANNA ZASIMOVA/Shutterstock; p121r Magicleaf/Shutterstock; p125t Maksim M/Shutterstock; p125cl Matsabe/Shutterstock; p125cr Bowrann/Shutterstock; p125b Vladvm/Shutterstock; p151l Dasha Petrenko/Shutterstock; p151cl Dancake/Shutterstock; p151ccl New Africa/Shutterstock; p151ccr GO DESIGN/Shutterstock; p151cr Tatyana Vyc/Shutterstock; p151r Albina Glisic/Shutterstock; p220t James Steidl/Shutterstock; p220b Nattiyapp/Shutterstock.